Laboratory Fume Hood Performance Testing Explained

AF816898

by

Chip Albright

FHC FUME HOOD CERTIFIED

A Fume Hood Certified Publication

Fume Hood Certified, LLC
PO Box 71477
Phoenix, AZ 85050 USA

www.fumehoodcertified.com

experts@fumehoodcertified.com

+1 623-662-0907

Disclaimer

This publication is designed to provide accurate and authoritative information with regard to the subject matter covered. It is sold with the understanding that the contents do not render professional advice. If you require professional assistance or expert advice, you should seek the services of a licensed professional.

Any company names, logos, or images used that are trademarked or copyrighted by others are acknowledged to be the property of the trademark or copyright owner.

This book was created with assistance from artificial intelligence resources. The content has been reviewed and edited by a human. For more information on the extent and nature of AI usage please contact Chip Albright.

Copyright © 2025 Fume Hood Certified, LLC

All rights reserved. No part of this book may be reproduced or used in any manner without the prior written permission of the copyright owner, except for the use of brief quotations in a book review. To request permissions, contact the publisher at

experts@fumehoodcertified.com

Paperback: ISBN **978-1-7357110-7-2**

First paperback edition December 2025.
Printed in the USA

Table of Contents

Introduction

In a forty–year career, you hear countless comments, debates, and side conversations. Most pass by like background noise. But every once in a while, a sentence lodges in your mind and refuses to leave. It changes how you see the world-and in my case, how I thought about fume hoods. Let me share three such moments.

The first was in 1999, at a SEFA-1 working summit. Picture the scene: a room filled with the industry's best minds-lab planners, manufacturers, engineers-wrestling with the wording of the new standard. The debate circled around an issue we just couldn't nail down. Then, almost casually, one well-known lab planner leaned back and said,

"The problem is, you can't see what the air is doing. If you could, this would all be common sense."

That hit me hard. Think about it, what you can't see can hurt you. An invisible enemy. The fume hood is supposed to be the holding cell for a dangerous criminal, but if you can't see the jailbreak and the fugitive chemicals, how do you know when it's happening?

A second moment came not long after, during an ASHRAE 110 committee meeting. A manufacturer reported unusual results when they tweaked the height of the SF_6 ejector. They were surprised by the results that changing the height made. Another member looked around the table and said bluntly,

"Why are you surprised? We don't design hoods to optimize containment. We design them to test well on ASHRAE 110."

That was another lightning bolt. Think about it: our most widely recognized performance metric was shaping design, not for real safety, but to pass a test. The focus had drifted away from building the most robust hoods possible and landed squarely on scoring well in the classroom of standards. Passing the exam had become more important than surviving the fire.

The third moment came about a decade ago in China. I had just finished a presentation when an international colleague approached me. He said,

"The reason more people don't test hoods is simple: it's too complicated and too expensive. Testing needs to be as easy as pushing a button and knowing if your hood works or not."

That was an aha moment. He was right. If testing is too complex or costly, people will avoid it-and that means too many labs are flying blind. Around the same time, I was teaching Lean Manufacturing and Visual Management, where one of the core principles is: "If you can't see a problem, you can't fix it." It brought me full circle to that first comment-if we could see loss of containment, the mystery would vanish. Everyone would understand, and fixing it would be a matter of course.

If you've followed me for a while, you've probably heard me say, "Fume hoods are the most misunderstood and misused safety device in the lab." Their testing follows right behind.

Remember-these devices are not furniture; they are exposure control devices. They are chemical containment devices. Which means the question is simple: are they containing, or are they not

This book is about that question. About containment. About the tests we use-and misuse-to answer it. In the chapters ahead, I'll walk you through the history, the standards, the myths, the evolving methods, and some ideas you may not have considered before. Along the way, I hope to challenge your assumptions, to make you question what you think you know, and to open your eyes to what testing should be: a way to validate real containment, not just to check a compliance box.

Settle in. Let's take this journey together.

About the Author

When **Chip Albright** walked into his first fume hood factory in the mid-1980s, he had no idea that he was stepping into what would become a lifelong mission. At Kewaunee Scientific in Lockhart, Texas, Chip was handed a clean sheet of paper and asked to help design better hoods. He poured himself into the work, quickly realizing that these weren't just pieces of steel and glass - they were life-saving devices, the first line of defense for countless scientists. That was the beginning of what would become a four-decade journey into the world of laboratory safety.

From those early days, Chip moved through the ranks of industry leaders like Jamestown-JMP and Collegedale LLC, always chasing the same question: How can we make fume hoods safer? Along the way, he designed seven major product lines and watched as hundreds of thousands of his hoods were installed in laboratories around the world. Each new project taught him something more about the delicate dance between hood design, laboratory ventilation systems, and the way real people actually use these devices.

But Chip didn't stop at design. In 1999–2000, as Chairman of SEFA, he led the rewrite of **SEFA-1: Fume Hoods**, a turning point in setting global standards. He became known not just as an engineer, but as a **voice of reason in an industry full of myths**, challenging the long-held belief that "face velocity" was the ultimate measure of safety. To Chip, it was always about one thing: containment.

That message began to resonate far beyond the United States. In **China and India**, Chip became a highly sought-after speaker and advisor. In India, his **Tri-Color Laser Testing workshops** drew crowds of EHS professionals and engineers eager to see - with their own eyes - how airflow really behaved inside a hood. In China, he worked alongside manufacturer, Sunway, to reimagine what a "smart hood" could be, helping to shape new generations of technology that were less dependent on perfect room conditions and more focused on user protection. Whether in Phoenix, Mumbai, or Shanghai, Chip's reputation has been the same: a straight-talking expert who can make the invisible visible and the complex understandable.

His book Laboratory Fume Hoods Explained and his many workshops, white papers, and presentations carry the same voice - practical, unvarnished, and deeply informed by years on factory floors, in labs, and at test benches. Through his company, **Fume Hood Certified LLC**, Chip has trained thousands, from lab managers to architects to young engineers, always reminding them that the hood is not just equipment - it's a lifeline.

Today, Chip's mission remains unchanged: **"Making labs safer, one fume hood at a time."** He continues to mentor the next generation, develop smarter testing tools like the **Fume Hood Performance Tracker**, and push for innovation wherever it's needed most. What drives him is simple - the belief that every laboratory worker deserves to go home safe, and that every hood should be trusted to do its job.

Chapter 1:
Busting the Myths and Misunderstanding

Myths and Misconceptions - What Everyone Gets Wrong About Fume Hoods

It always starts with a number. "100 feet per minute."

Ask a group of lab managers, safety officers, or even architects what that phrase means, and they'll nod knowingly. For decades, 100 fpm has been the magic mantra, the supposed silver bullet of fume hood safety. If the gauge says the hood is drawing 100 fpm at the sash opening, then the scientists are safe. End of discussion. Or so everyone thinks.

But what if I told you that this belief-this cherished metric-isn't the guarantee of protection you imagine it to be? What if it's not only incomplete but potentially dangerous? What if your focus on that number is putting people in the lab at risk? Many of these myths we will explore are like urban legend, that generally are rooted in history but as they are passed along, they take on a life of their own. In the book we will explore some of the history that allows these myths to exist.

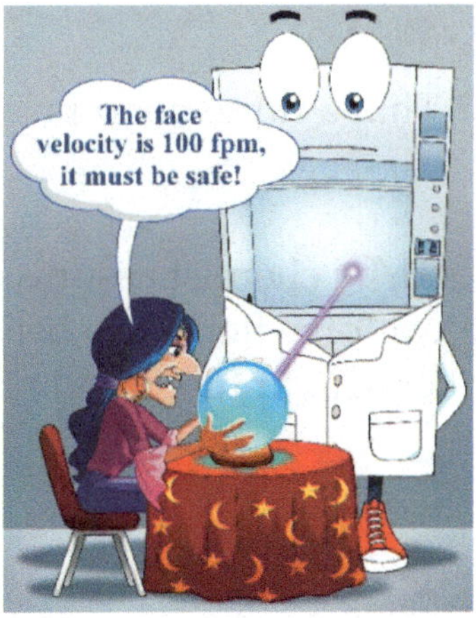

The Myth of the Magic Number

Picture a graduate student, Sarah, working late at night. She checks the hood monitor: 101 fpm. Satisfied, she leans in close to adjust her beaker. She believes the number glowing green means containment is assured. Yet tracer gas studies and modern visual testing (like the Tri-Color laser) have repeatedly shown that even at "perfect" velocities, contaminants can stream back into the breathing zone if the airflow is uneven or turbulent. How would the user know if the performance of their hood was really safe or not? Blind trust! If there is a hood in my lab, it must be safe.

Here's the uncomfortable truth: **face velocity is about airflow quantity, not quality.**

Air doesn't move like soldiers marching in neat rows; it moves like a river-sometimes smooth, often swirling with eddies, turbulence, and cross-currents. Just as water can escape a riverbank through a whirlpool, chemical vapors can escape a fume hood through chaotic airflow patterns.

In fact, research has shown that excessively high face velocities-those above 130 fpm-can create so much turbulence that containment worsens. So much for "more is better."

The "Absolute Protection" Myth

Another dangerous misconception? That a fume hood provides total protection.

"Don't worry, I'm doing it in the hood," a chemist assures a colleague, as if the sash were a magical force field. The belief is that all chemicals in the fume chamber are captured and contained and, all risk vanishes.

But the reality is different. **Hoods are not absolute containment devices; they are exposure reduction devices.** Even the best-tested hoods may leak trace amounts, and improper work practices-placing equipment too close to the sash, moving rapidly, or propping the sash wide open-can turn that "protection" into an illusion. OSHA is clear: a hood is both an engineering control and a form of PPE, not a guarantee of zero exposure. While a fume hood is an exposure control device, and preventing exposure it the primary purpose for operating fume hoods, nothing is perfect.

The "If I Don't Smell It, It's Safe" Myth
Let's tackle another: the nose as a safety device. "If I can't smell it, it must not be leaking."

This logic is as flawed as assuming that if you don't hear your car engine knocking, it must be running perfectly. Many hazardous vapors-benzene, carbon monoxide, formaldehyde-either have no odor at dangerous concentrations or only become detectable after exposure exceeds safe limits. Relying on smell is not only misguided, it's reckless.

The One-and-Done Myth
Then there's the assumption that once a hood has passed certification, it will forever be safe. Imagine if pilots tested their aircraft only once, at delivery, and never checked it again. Unthinkable, right? Yet labs often assume a passed test is a permanent seal of safety. But hoods live in dynamic ecosystems. A new supply diffuser installed in the ceiling, a door propped open down the hall, or a technician who prefers working with the sash raised higher than designed; each can change airflow and cause containment loss. That's why standards like ASHRAE 110 emphasize "as used" testing.

A hood that worked yesterday may not work tomorrow if the environment changes. And room conditions are anything but perfect. They are ever changing which means the hoods performance is ever changing

Loss of Containment: The Silent Threat

Have you ever seen loss of containment? Most people haven't-because it's usually invisible. Yet laser visualization shows us how often there is loss of containment. Studies suggest that nearly half of hoods would fail containment testing even when their face velocity is within spec. The causes break down like this:

- 25% from hood or room design flaws
- 50% from room conditions and ventilation systems
- 25% from user behavior and setup

The majority of the failures are not about the hood itself at all-but about where it is and how it's used.

You bought the hood and had it installed and tested. Now you are doing some form of regular testing, what are you testing? The hood didn't change nor did its location. Therefore, performance has little to do with the hood itself, it about room conditions and user practices.

What Level of Loss Is "Safe"?

This question makes people uncomfortable, but it must be asked: Is any leakage acceptable? The answer depends on context. A tiny leak of a relatively benign solvent may pose little risk, while the same leak of a carcinogen or highly toxic gas may be catastrophic. Standards like ANSI Z9.5 and SEFA stress that risk assessment must always be chemical-specific. There is no blanket assurance.

The Face Velocity Trap

If face velocity isn't the yardstick, why does the myth persist?

Because it's easy. It gives a simple number to hang your hat on, a quick box to check. But science is rarely that simple. A single average reading at one moment in time tells us about as much as a snapshot of freeway traffic tells us about congestion trends over a year.

Real containment is about dynamic performance, not static numbers. That's why the industry is moving toward visualization and tracer gas testing-tools that reveal the real story, not just the illusion.

Changing the Narrative

In truth, every myth we've discussed has a common root: the human need for certainty. We want the hood to be a magic box. We want numbers to mean safety. We want one test to last forever. But reality is more complex, and that complexity demands humility and vigilance.

The real lesson-

Containment is not guaranteed by a number, a test, or a piece of equipment-it's the product of design, environment, and behavior working together.

When we begin to ask the hard questions-what level of loss is acceptable, how do my practices impact airflow, what does this test actually prove-we start to replace myths with understanding. And only then do we move from blind trust to informed safety.

Why Is It That We Test?

If you ask a room of laboratory managers or EHS professionals why they test their fume hoods, you'll hear a chorus of answers:

- "It's a requirement of our organization."
- "It's a compliance issue."

- "Our accrediting agency requires it."
- "It's a government requirement."
- "To mitigate risk if there's an accident."
- "To make the labs safer."
- "To protect the occupants."
- "To minimize the risk of exposure."

All of those reasons are true. But if you step back, none of them really capture the heart of the matter.

The reason we test is much simpler-and far more important. We test to validate that a hood is performing safely in the room it lives in, under the conditions it faces, and with the people who actually use it.

The Trouble with Assumptions

Too often, people assume that if a fume hood was designed well and installed correctly, it must be working. Or they trust the sticker on the sash, confident that the average face velocity number written there is a guarantee of safety. But air is invisible, dynamic, and deceptive.

A hood that once performed flawlessly can lose containment without warning. It might be because the building's HVAC system switched from heating to cooling mode, altering air currents in subtle but significant ways. In heating mode, warm air hugs the ceiling and drifts downward as it cools. In cooling mode, cold air plunges directly to the floor. Those patterns are opposite-and if the supply diffuser is poorly placed, the crossdraft created can literally pull contaminants out of the hood. Add strong sunshine across the bench, and you've introduced another invisible disruptor.

Even something as ordinary as a user placing a hotplate inside the hood can cause a shift. Rising heat creates convection currents that collide with the designed airflow. Suddenly, vapors that should have been captured and exhausted are spilling back toward the breathing zone.

Why Face Velocity Isn't Enough

For decades, the industry clung to a myth: 100 fpm face velocity equaled safety. But research and countless field tests have shown otherwise. A hood can read "perfect" on the meter and still leak contaminants. Conversely, a hood at 60 fpm can perform beautifully if the airflow is smooth and undisturbed.

Face velocity is just a snapshot. It measures how much air is being pulled through the sash at that moment, but it doesn't measure containment. And containment-the prevention of fugitive vapors escaping into the lab-is the real measure of safety.

This is why ASHRAE 110 and its derivatives added tracer gas testing and flow visualization. Tools like Tri-Color laser testing take it a step further, letting users actually see the invisible currents of air-and, sometimes, the heartbreaking failures when vapors escape.

One Size Does Not Fit All

So how often should a hood be tested? The answer depends on risk. A teaching lab handling dilute acids might only need annual certification. A pharmaceutical lab working with cytotoxins might need quarterly-or even post-change-validation. Every move of equipment, every adjustment of a sash stop, every new process brings with it the possibility of changing airflow patterns.

Standards like ANSI Z9.5 and ASHRAE 110 emphasize this principle: testing is not a box to check; it's a management tool for exposure risk.

The Real Motivation

Ultimately, the correct motivation for testing is not compliance, not paperwork, and not even liability protection. It's prevention. We test to minimize the chance of a lab accident. We test to prevent occupants from breathing what they cannot see.

And here's a sobering truth: a poor test may be worse than no test at all. A false "pass" provides a dangerous illusion of safety, lulling users into risky behavior. Real testing-dynamic, thoughtful, risk-based testing-pulls back the curtain and tells us the truth.

A Story to Remember

I once witnessed a hood in a university chemistry lab that had passed its annual face velocity check at 105 fpm. The sticker was proudly displayed, and the users trusted it. But during a containment visualization test, the Tri-Color laser revealed green beams spilling straight out of the sash into the breathing zone. Nobody could believe it-least of all the professor whose students had been relying on it for months.

That's why we test. Because the stakes are invisible until they're not.

By the end of this book, my goal is that you'll be able to answer the question "What is the appropriate metric for safe fume hood performance?" with more clarity and confidence than you had when you started.

There is no magic number. No single, universal answer. Only shades of gray, informed judgment, and the discipline of testing until you know.

.Chapter 2:
The History of Fume Hood Performance Testing Methods

If you step back and look at the history of fume hood testing, one theme jumps out: the standards have always lagged behind reality. Long before there were formalized procedures, laboratories had fume hoods in place and needed some way to decide if they were safe. What did they do? They measured air speed at the sash opening. If the hood "pulled" air at roughly 100 feet per minute, it was declared safe. That simple rule of thumb became the de facto benchmark, even though-as we've known for decades-it told us little about whether the hood actually contained hazardous fumes.

By the late 1970s, pioneers like Caplan and Knutson were already questioning that approach. They began experimenting with tracer gases, asking a more direct question: not how fast the air was moving, but whether contaminants were escaping. Their work planted the seed for what would become the modern era of performance testing.

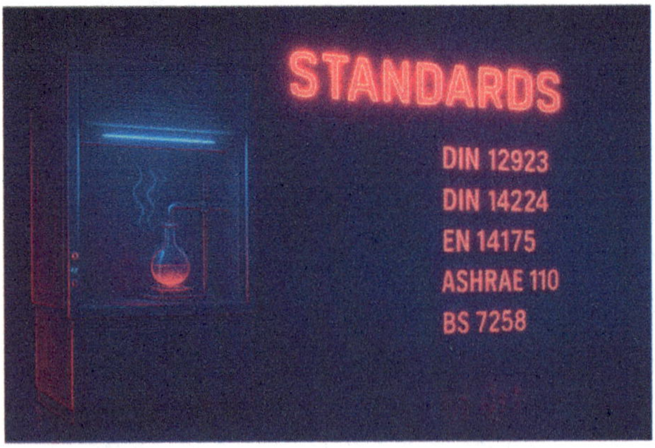

Germany Led the Way
Germany was first to codify laboratory hood standards. DIN 12923 appeared in 1975, followed by DIN 12924 in 1978. These early efforts were still rooted in airflow metrics, but by 1991 Germany had made the leap to tracer gas containment testing. For the first time, a standard didn't just measure how air moved-it measured whether dangerous vapors stayed inside the hood.
That shift from airflow volume to containment performance was revolutionary, and it laid the foundation for Europe's later EN 14175 standards.

The American Breakthrough - ASHRAE 110 (1985)
Meanwhile, in the U.S., the pressure was mounting for a comprehensive test method. In 1985, ASHRAE published Standard 110, the first dedicated fume hood performance standard anywhere in the world. It was groundbreaking because it combined three approaches into one test:

- **Face velocity profiles** to map airflow across the opening,
- **Smoke visualization** to see flow patterns and turbulence,
- **Tracer gas containment** to directly measure whether contaminants reached the operator's breathing zone.

That last piece-the tracer gas test-was the real breakthrough. For the first time, we could put a number on how safe a hood really was. Not just "it feels like it's pulling air," but a quantifiable measurement of containment. ASHRAE 110 quickly became the benchmark not only in the U.S. but around the world.

Britain's BS 7258 (1990s)
Across the Atlantic, the British Standards Institution introduced BS 7258 around 1990. Like Germany, they started with face velocity traverses but soon moved toward containment testing. By 1994, Part 4 of BS 7258 formally incorporated tracer gas methods, aligning the UK with the same performance-based philosophy that ASHRAE had set in motion. The push in Britain was accelerated by regulatory pressure-COSHH (Control of Substances Hazardous to Health, 1988) required rigorous checks of fume cupboards, and BS 7258 provided the way to do it.

Converging Paths
By the 1990s, the U.S., Germany, and the UK had all arrived at the same conclusion: face velocity alone could not guarantee safety. Containment had to be tested directly. From there, harmonization followed-EN 14175 in Europe, revisions of ASHRAE 110 in 1995 and 2016, and the integration of tracer gas methods into other standards like ANSI Z9.5.

The Long Shadow of ASHRAE 110
Here's the irony: ASHRAE 110 was visionary in 1985, but the standard itself has barely changed since. The 2016 update added some digital data requirements and clarifications, but the core methodology is the same as it was 40 years ago. The committee recently reaffirmed it again, meaning it may remain essentially unchanged until 2036-a full half-century after its debut. In that same span, hood designs, building controls, and visualization technologies have advanced dramatically. Meanwhile, the field has innovated on its own. Tri-Color laser visualization, real-time diagnostics, and performance tracking tools have become powerful ways to evaluate hoods under dynamic, real-world conditions. These methods aren't (yet) enshrined in standards, but they are rapidly becoming the de facto standard in labs that take safety seriously.

When Standards Lag, Innovation Leads

That is the story of fume hood testing. The standards give us a baseline, a common language. But they are slow to change-tied to consensus processes and 10-year review cycles. Reality doesn't wait. Laboratories face new risks, technologies evolve, and innovators push forward with tools that quickly become accepted practice long before the standards catch up.

So, the lesson from history is this: if you want to understand where fume hood performance testing is going, don't just read the standards. Watch the innovators. That's where tomorrow's "standard practice" is already being written.

Chapter 3:
What ASHRAE 110 Is - and Isn't

If you spend enough time around laboratory people, you'll hear the phrase "ASHRAE 110" spoken with the weight of gospel. It is quoted in specifications, referenced in compliance reports, and quite often totally misunderstood. Some believe it tells us if a hood is "safe." Others treat it as a pass/fail exam. The truth is more nuanced.

To understand why ASHRAE 110 exists at all, we need to step back into the late 1970s and early 1980s. At that time, there was no universal method of testing fume hoods. Manufacturers competed fiercely, each claiming their hood was "the best in the world." Buyers had no reliable way to compare one model to another. Most fume hoods were designed and sold by furniture companies-treated as cabinetry rather than sophisticated ventilation devices. Something had to change.

The Birth of a Standard
Enter ASHRAE 110, first published in 1985. The committee's goal wasn't to define "safe" operation but to create a reproducible method of testing. Think of it less as a referee declaring winners and losers, and more like a stopwatch-providing standardized numbers that let you compare performance.

The test protocol was simple but powerful. It called for a "perfect" test room-free from the drafts, door openings, and people walking by that plague real laboratories. Within this idealized room, three things happened:

1. **Face velocity profiling** - measuring airflow across a grid at the sash opening.

2. **Smoke visualization** – using smoke to make the invisible visible.

3. **Tracer gas containment** – releasing sulfur hexafluoride (SF_6) and measuring how much leaked out.

The results weren't meant to say "this hood is safe" or "this hood is unsafe." They were designed to give buyers comparable data across manufacturers.

A Testing Protocol, Not a Safety Standard
This distinction is critical. ASHRAE 110 itself is explicit:
This standard defines a reproducible method of testing laboratory hoods. **It does not define safe performance.**
That's why ASHRAE 110 is:

- A **testing protocol**, not a safety standard.
- **Not pass/fail Standard**
- **Not a direct measure of safety.**
- A **snapshot in time**, under simulated conditions-not the messy reality of a working lab.

ANSI Z9.5, the broader Laboratory Ventilation Standard, goes a step further by suggesting benchmarks for interpreting tracer gas results: for example, 0.05 ppm as acceptable in the "As Manufactured" test, and 0.10 ppm in "As Installed" or "As Used" conditions. But even there, the documents are careful to warn: containment in a test chamber is not the same as safety in real life.

The Fine Print That Matters
Buried in the language of ASHRAE 110 are truths many forget:

- Tracer gas doesn't simulate real chemicals, their risks, or how users interact with a hood.
- A mannequin isn't a live chemist. Real people lean in, move quickly, block air paths.
- Passing the AM (As Manufactured) test in a factory doesn't guarantee the hood will protect users in the AU (As Used) conditions of a busy laboratory.

Even the committee itself warned:
"Values for control level may not be suitable for establishing hood safety... the tracer gas test does not simulate a live operator." In other words: don't mistake the yardstick for the race.

Where the Confusion Creeps In
Why do so many believe ASHRAE 110 is a safety stamp of approval? Because it produces numbers-and humans like numbers. A report that says "0.03 ppm" feels comforting, like a lab test coming back negative. But those numbers are only meaningful in context. A hood might ace its factory AM test and still perform miserably once installed in a lab with swirling cross-drafts, a badly balanced HVAC system, or poor user practices.

That's why newer methods like **Tri-Color Laser Visualization** have gained traction. They don't replace ASHRAE 110, but they help fill its gaps-showing in real time how containment behaves when someone walks by, opens a door, or waves a hand near the sash.

The Bottom Line

So, what is ASHRAE 110? It is a tool. A protocol. A way to measure hoods consistently across manufacturers and installations. It gives us data points, not guarantees.

And what isn't it? It is not a declaration of safety. It is not a substitute for risk assessment, user training, or periodic real-world testing.

The title of the standard says it all: **"Methods of Testing Performance of Laboratory Fume Hoods."** It tells us how to test-not what is safe.

The responsibility for defining safety lies not in the standard, but in the cooperative efforts of users, safety officers, ventilation engineers, and those of us who keep asking the uncomfortable question: Is this hood really protecting the person standing in front of it?

Let the standard speak for itself. Let's look at exactly what ASHRAE 110 - 2016 actually says:

"This standard defines a reproducible method of testing laboratory hoods. It does not define safe procedures. However, laboratory hoods are considered by many to be the primary safety devices in conducting laboratory operations."

"There are many important factors in the safe operation of laboratory hoods that are not described in this standard."

"ANSI/ASHRAE 110 may be used in the laboratory as an accepted test with specific values for the control levels (and the release rate if you depart from the standard). It also may be used for routine periodic testing, but it is somewhat expensive and other less rigorous tests may be adequate if conditions have not changed since commissioning tests."

"The test data should be available from the manufacturer for each specific model and type of hood so a potential buyer to assist with design specifications or compare one manufacturer's hood containment against another."

"Values for control level may not be suitable for establishing hood safety, as the tracer gas test methods may not adequately simulate actual material use, risk, or generation characteristics. In addition, the tracer gas test does not simulate a live operator, who may increase potential for escape due to operator size, movements near the hood opening, or improper hood use."

"Hood containment should be evaluated at different mannequin heights to represent workers of different height. AM 0.05 can be achieved with a properly designed laboratory fume hood. It should not be implied that this exposure level is safe. Safe exposure levels are application specific and should be evaluated by properly trained personnel (SEFA 1-2010, Recommended Practices for Laboratory Fume Hoods)"

"The tracer gas, Sulfur Hexafluoride (SF6), typically used during ANSI/ASHRAE 110 fume hood performance tests has been identified as a very potent Green House Gas (GHG). Generation of SF6 is discouraged and banned in some locations requiring specification and use of alternative tracers. ASHRAE has undertaken an investigation to evaluate alternative tracers but have not approved a more environmentally friendly replacement of SF6. Where SF6 tests are discouraged or banned, alternative methods can be utilized where approved by the owner, responsible person and the design team."

"In summary, there are many factors to consider in evaluating the performance of a laboratory hood installation. This standard provides one tool in evaluating such safety"

“The procedure is a performance test method and does not constitute a performance specification.”

“The desired hood performance should be defined through the cooperative efforts of the user, the chemical hygiene officer, the applications engineer, and other parties affected by the hood performance. It should be noted that the performance test method does not give a direct correlation between testing with a tracer gas and operator exposures.”

“The performance test method does, however, give a relative and quantitative determination of the efficiency of the hood containment under a set of specific, although arbitrary, conditions.

The same test can be used to evaluate hoods in manufacturer facilities under (presumably) ideal conditions or under some specified condition of room air supply or during the commissioning of a new or renovated laboratory before the user has occupied the laboratory”

“The full test procedure (visualization, face velocity, and tracer gas) is a quantitative measurement of a hood’s containment ability and is useful for hood development and rigorous evaluation of hood performance.”

“This standard defines a reproducible method of testing laboratory hoods. It does not define safe procedures. However, laboratory hoods are considered by many to be the primary safety devices in conducting laboratory operations”

“It is important to evaluate the performance of the laboratory hood under dynamic conditions. This performance test method may be modified_to evaluate a dynamic challenge.”

“This test method addresses only the dynamic challenge of sash movement.”
“VAV hoods place a significant emphasis on the sash movement and the potential effect on hood performance. However, some constant-volume hoods may also experience a decrease in protection when the sash is moved.”

Based on these excerpts from the ASHRAE 110 - 2016 Standard what do you think the standard is?

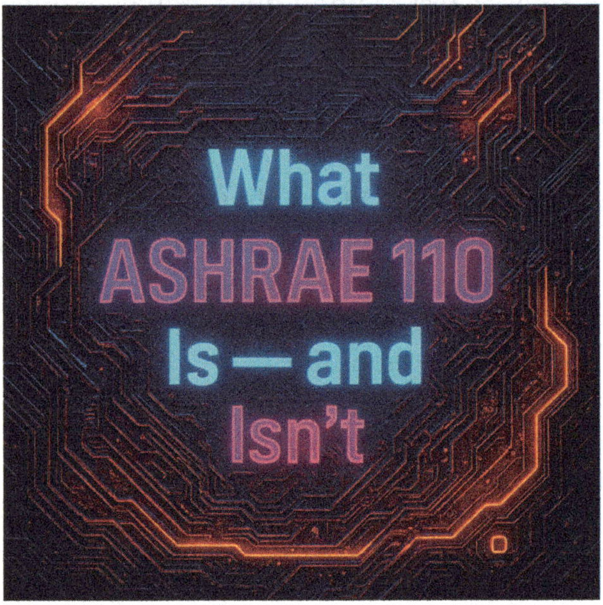

While it may be boring, the best way to really understand ASHRAE 110 is to read it word for word. I not only did that, but I highlighted key information and have it laminated and take it with me whenever I do field testing.

Chapter 4:
ANSI/ASSP Z9.5 - 2022: The Bigger Picture

If ASHRAE 110 is a microscope that zooms in on the fume hood itself, Z9.5 is a wide-angle lens. It looks at the entire laboratory ventilation system-the supply, the exhaust, the controls, and the management plan that ties it all together. Because the truth is simple: a fume hood can't protect anyone until it's connected to a properly designed and maintained system.

Back in the early days, people often treated fume hoods like magic boxes-pulling bad air out simply because someone said they did. Lots of smoke and mirrors. But real containment depends on the invisible river of air flowing through the lab. Z9.5 is the standard that tries to bring order to that invisible world.

Think of air like water. Water flows downhill; air flows from high pressure to low pressure. When someone says, "this hood sucks," what they're really describing is that pressure difference. The hood's chamber becomes a low-pressure zone, and air is pulled in to balance it out. Simple in principle, but endlessly complicated in practice.

Z9.5 recognizes that complexity. It doesn't just ask, "How fast is the air moving at the face of the hood?" Instead, it asks bigger questions:

- Is there a written management plan for how the lab's air will be controlled?
- Is someone responsible for making sure the system is maintained and tested?
- Are risks being assessed before new chemicals, new processes, or new equipment are introduced?

- Are users trained not just to trust the hood, but to understand its limitations?

The standard reminds us that safety isn't just about a number on a meter. Face velocity, for example, is a useful piece of data, but not the whole story. Too little airflow and fumes can escape; too much and turbulence can actually pull contaminants out toward the user. Containment is about balance, and that balance is influenced by everything from sash height to cross-drafts from an open door.

Z9.5 also makes a key point that often gets overlooked: fume hoods are exposure control devices, but they aren't the right answer for every hazard. Some chemicals require gloveboxes, special-purpose hoods, or even unique ventilation designs. Another major theme is accountability. Z9.5 requires labs to have a Laboratory Ventilation Management Plan (LVMP)-a living document that spells out how the ventilation system will be designed, operated, and reviewed. Someone must be clearly named as "responsible person" for it. Without that accountability, labs drift into a dangerous assumption that "the hood will take care of it."

The standard even looks at energy use, recognizing that ventilation systems are one of the largest energy consumers in a lab building. It encourages designs that protect people and reduce wasted energy, because sustainability and safety are no longer separate conversations.

At its heart, Z9.5 tells us this: laboratory ventilation is not just a piece of equipment, it's a system. And like any system, it only works when all the parts-design, maintenance, training, and management-come together. We need a holistic view.

So, while ASHRAE 110 helps us test the fume hood in a controlled way, Z9.5 reminds us to step back and see the whole forest, not just the tree. If the ventilation system isn't right, the hood won't be right either.

One of the places where Z9.5 brushes up against ASHRAE 110 is on the subject of leak rates. The standard references an "acceptable" tracer gas escape rate-often pegged around 0.10 ppm. That sounds clear-cut, but it immediately raises an uncomfortable question: why is 0.10 considered safe while 0.11 is considered a failure? The truth is, safety can't be wrapped up so neatly. Airflow in a lab is dynamic, and containment is influenced by countless small factors-room drafts, sash position, user movement. The number is less of a hard line in the sand and more of a guidepost, reminding us that testing containment is about managing risk, not chasing perfection.

Face velocity falls into a similar category. For decades, 100 feet per minute was treated as the "magic number" for hood safety, and many labs still cling to it. Z9.5 softens that stance. It acknowledges that face velocity is a useful measurement, but not a guarantee of containment. Too low, and contaminants may not be drawn in. Too high, and turbulence can actually push fumes out toward the user.

That's why the standard recommends a range-typically 80–100 fpm-as a balance point. More importantly, it emphasizes that velocity is only part of the story. The real measure of a hood's performance is whether it keeps contaminants inside, and that requires looking at the whole system, not just the speed of the air at the sash.

Chapter 5:
Why ASHRAE 110 Has Three Versions

Most people who first hear about ASHRAE 110 testing wonder the same thing: Why three versions? Isn't one test enough to tell you if a hood is safe?

The truth is, each of the three versions-AM (As Manufactured), AI (As Installed), and AU (As Used)-captures a very different reality. They are not redundant; they are sequential chapters in the hood's life story. To understand why, we need to step back and look at the logic behind them.

AM - The Glamour Shot

Every hood model begins its life in a controlled test room, undergoing the **AM (As Manufactured)** test. This is the hood's glamour shot-perfect lighting, no distractions, a carefully prepared stage. Manufacturers build special test rooms as specified in the ASHRAE 110 standard designed to minimize variables. The airflows are tuned, cross-drafts are reduced, and the room is arranged to give the hood its best chance to succeed.

The purpose is fairness. Imagine trying to choose a car if each automaker was allowed to test on different tracks-some smooth and flat, others full of potholes. You could never make an apples-to-apples comparison. The AM test solves that by leveling the playing field. It allows specifiers and buyers to compare performance data between hood models under identical conditions.

But there's a catch. AM results often make a good hood and a great hood look the same. Both may score nearly identical containment numbers. Yet in the real world, the great hood will outperform the good one when the lab environment is less than perfect.
That's why AM is essential, but incomplete. It shows us what a hood can do in an ideal world, but it doesn't tell us how it will behave when things get messy.

AI - The First Day of School
The next chapter is **AI (As Installed)**. Picture the hood arriving in its new home, freshly uncrated and connected to the building's exhaust system. The ductwork is sealed, the fans are humming, and the supply air is balanced. Nobody has used it yet. Everything is shiny and new.

AI testing is like the hood's first day of school. It's no longer posing for the studio portrait-it's walking into a classroom full of other kids, learning how to fit in. The test validates that the hood works properly with its new environment. It checks that the ventilation system is balanced, the room air supply is cooperating, and the hood is performing as designed.

This is also where we compare the AI data to the AM report. If the results line up, the hood and the room are in harmony. If they don't, the problem isn't the hood-it's the room conditions. The hood itself hasn't changed since the factory. AI provides a crucial benchmark, a baseline for performance in the actual lab environment.

Unfortunately, many labs skip or downplay AI testing. They assume that if the hood passed AM at the factory, it must be fine now. But anyone who has commissioned labs knows that's a dangerous assumption. The exhaust fan might be undersized. The supply air diffusers might create turbulence at the sash. A beautifully performing hood in AM can stumble badly once installed.

Without AI testing, you're essentially sending your child off to school without checking if the bus even comes to your neighborhood.

AU - Real Life, Messy Life

Finally, we arrive at **AU (As Used)**, the test that matters most to the people actually working at the bench. This is where the hood is evaluated under real-world conditions-when the lab is occupied, the benches are cluttered, other hoods are in operation, doors are opening and closing, and people are walking past.

If AM is the glamour shot and AI is the first-day-of-school picture, AU is the candid snapshot taken years later-backpacks on the floor, messy hair, and real life happening in the background. It's not always pretty, but it's real.

AU testing is often eye-opening. Many testers have a story about being called in to conduct AU tests, only to find the hoods "cleaned out" by lab staff. The benches are spotless, the glassware neatly stowed, and the hood looks more like a museum display than a working piece of equipment. That might make the test easier, but it misses the point. Nobody works in an empty hood. Real AU testing should challenge the hood as it is being used-with equipment, clutter, and activity included.

Done properly, AU reveals the truth. It shows how containment holds up when turbulence is created by human activity, when cross-drafts occur, and when the unpredictable rhythm of a busy lab collides with the engineered precision of the ventilation system.

The Whole Story

Taken together, the three tests form a progression:

- **AM** - What the hood can do in a perfect environment.
- **AI** - How the hood behaves when connected to its new lab.
- **AU** - How the hood performs in the chaos of daily use.

An empty hood in a working lab is neither an AI or AU

It's tempting to rely on just one test, but doing so is like trying to understand a child's development from a single photo. At one year, six years, and eighteen years, the same child looks completely different. Each snapshot is incomplete by itself, but together they tell the story of growth and change.

The same is true for fume hoods. Without AM data, you can't compare models. Without AI data, you can't verify proper installation. Without AU data, you can't know if your hood is actually keeping users safe.

Lessons from Other Industries

This three-stage approach isn't unique to fume hoods. The auto industry learned a similar lesson decades ago. For years, crash testing focused almost exclusively on frontal impacts. Cars were designed to ace those tests, and consumers trusted the star ratings. But when side-impact testing was added, many "five-star" vehicles suddenly looked vulnerable. Real-world accidents had exposed weaknesses that the original test never revealed.

The parallel is clear. If you only test hoods under one condition, you risk being blindsided when reality introduces new variables. Safety isn't about passing a single test; it's about building resilience across multiple scenarios.

Common Missteps

Over the years, I've seen the same mistakes repeated:

- **Relying only on AM data.** Buyers sometimes assume a hood that performed well at the factory will automatically be safe in their lab. That's like assuming a race car built for the track will handle fine in rush-hour traffic.

- **Skipping AI commissioning.** Some projects rush to occupancy without proper AI testing. The hoods may look fine, but they've never been validated in their new environment.

- **"Sanitized" AU tests.** Labs often clear out clutter before AU tests, creating artificial conditions. It makes the numbers look better, but it tells you nothing about actual risk.

Each of these shortcuts undermines safety. Testing is only meaningful when it reflects reality.

Building the Hood's Baby Book

The best labs I've worked with treat hood testing as a continuous story. They keep records of the AM test provided by the manufacturer, the AI results at commissioning, and the AU data from annual or semi-annual testing. Over time, this creates a "baby book" for each hood-a chronological record of its performance through the many years of operation.

With that history, you can see trends. You can spot a hood that is slowly losing containment. You can compare performance before and after a renovation. You can identify whether failures are due to the hood itself, the ventilation system, or user behavior.

Just as a child's baby book preserves milestones and growth, a hood's test history preserves its safety story. And just like a parent flipping through old photos, a lab manager can look back and see how the hood has changed over time-and whether it's still providing the protection it was designed to give.

The Bottom Line
ASHRAE 110 isn't complicated for the sake of complication. The three parts exist because no single test can capture the full reality of fume hood performance.

- **AM** shows the hood's maximum potential.
- **AI** validates its integration with the lab.
- **AU** proves its ability to protect in daily use.

Together, they tell the whole truth.

The next time someone asks why there are three versions of the test, tell them this: It's not about redundancy. It's about reality. Just like life, a fume hood's story can't be told with a single snapshot. It takes a series of moments, documented faithfully, to reveal whether that hood is truly keeping its users safe.

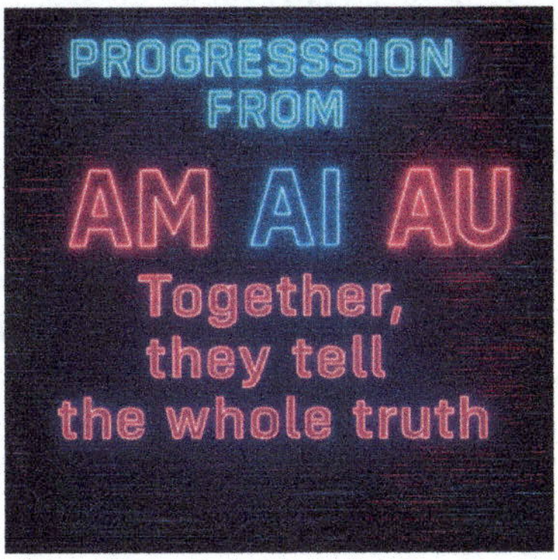

Chapter 6:
What is a Face Velocity Profile?

If you asked me to pick just one test from the ASHRAE 110 protocol-the one test I wouldn't leave the room without-it would be the **Face Velocity Profile**. This is the test that speaks the loudest, if you know how to listen.

Most people, when they hear "face velocity," think only about a single number: 100 feet per minute. For decades, that has been the magic number. But containment is not about hitting 100. It's about the quality of airflow, not the quantity alone. And nothing reveals the quality of airflow across the hood opening better than the face velocity profile.

Why This Test Matters
The arch-enemy of containment is turbulence. Think about a river: water can flow smoothly and calmly downstream, or it can churn into rapids. Both have flow, both have speed-but only one is safe to paddle. The same holds true for air entering a fume hood. A profile test shows us if the flow is smooth and laminar-or broken and turbulent, the kind of air movement that lets vapors escape into the room.

The profile test gives you a map of the invisible. It shows you not just the average face velocity, but whether that velocity is stable, uniform, and predictable-or a patchwork of highs, lows, and swirling eddies. A hood with turbulence is like a prison door with a bent lock: you might not notice the flaw until the prisoner is already out.

How the Test is Done
Imagine drawing an invisible checkerboard across the sash opening. Each square-called a cell-must be no larger than 12 by 12 inches (ASHRAE says 13, but I like smaller). The more cells you measure, the sharper your picture of the hood becomes.

You start at the first cell. A probe from a hotwire anemometer sits at the center. Connected to a data logger, it records one velocity reading every second for 20 seconds. You average those numbers to get the cell's velocity. Then you move to the next square, and the next, until the entire sash opening has been measured.

Once the grid is complete, you calculate the overall average face velocity. Then comes the real insight: each cell should fall within ±10% of that average. Record the highest cell and the lowest. The spread between them tells you the story of turbulence.

In a factory test room-the AM test environment where everything is perfect-we strive for ±2%. That's nearly laminar, the holy grail of hood aerodynamics. But in the real world, ±10% is considered acceptable. Anything wider means turbulence is lurking.

Why It Tells You More Than a Snapshot

One quick reading at the center of the sash? That's a snapshot-like guessing traffic speed by watching one car go by. It tells you something, but not enough. A profile test is a time-lapse movie. You see not only the average, but the highs and lows, the stable and the erratic.

I often expand the test beyond the standard. For field testing, I'll hold the probe in each cell for 90 seconds. For serious troubleshooting, I'll stretch it to 300 seconds-the same duration as the tracer gas containment test. When I do that, I'll run the SF_6 containment test and the velocity test side by side. The correlation is striking. Areas with turbulence on the profile are the very places where tracer gas escapes.

The Profile as a Story of Containment
Here's the real truth: face velocity alone does not guarantee containment. In fact, many hoods that fail tracer gas containment tests show perfectly "good" face velocities. What failed wasn't the average-it was the turbulence. The profile reveals that turbulence before the tracer gas ever does.

Think of the profile as a fingerprint of the hood's breathing. A clean, even profile means the hood inhales smoothly, pulling vapors back into itself. A jagged, uneven profile means the hood is gasping, creating whirlpools that can spit contaminants into the lab.

Lessons Learned in the Field
I've tested thousands of hoods, and the face velocity profile has never lied to me. Once, in a university chemistry lab, the average velocity looked textbook: 95 fpm. But the profile told a darker story. The upper-left corner cells were running 40 fpm higher than the average. The turbulence in that region lined up perfectly with where a student had placed her solvent bottles. When we ran the tracer gas, the leaks showed up exactly there. The average number had hidden the danger. The profile revealed it.

Another time, in a pharmaceutical lab, a hood kept failing SF_6 containment even after multiple balance adjustments. The profile showed the culprit: wild swings in velocity at the bottom edge. The room's supply diffuser was aimed right at the sash opening, creating a pulsing draft that not only fooled the anemometer but pushed contaminants into the breathing zone. Fixing the diffuser-not the hood-solved the problem. Without the profile, we would have chased ghosts for days.

Beyond the Standard: Modern Enhancements
ASHRAE 110 requires this test, but even ASHRAE admits it is just the start. Today, many testers, myself included, enhance the process with visualization. Tri-Color laser testing, for example, can be run alongside the velocity profile. Where the profile numbers suggest turbulence, the lasers make it visible: shimmering curtains of air spilling from the sash.

It's also worth noting that Z9.5, the ventilation system standard, echoes the importance of airflow stability across the hood face. It acknowledges that acceptable containment is not a single velocity number but the ability of the system to maintain safe, uniform airflow under varying conditions.

The Bottom Line
If you take away only one lesson, let it be this: **a face velocity profile is not about chasing a number, it's about uncovering turbulence.**

Numbers can lie. Averages can mislead. But turbulence always tells the truth. And the face velocity profile is our best tool to find it.

When I walk away from a hood, I want to know-not guess-that its breathing is steady, its flow is smooth, and its turbulence is under control. That's why, if I could only run one test, the face velocity profile would always be the one.

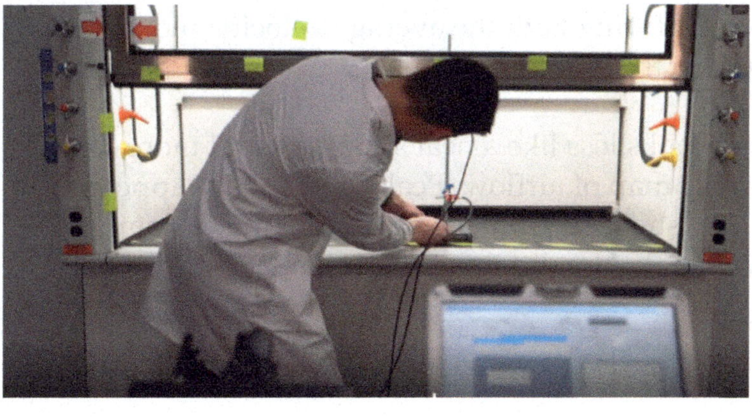

Chapter 7:
How does a Face Velocity Profile differ from Average Face Velocity?

The Illusion of the Average

Walk into any laboratory, glance at a certification sticker on a fume hood, and you will often see a single number proudly displayed: 100 feet per minute. That number-the **average face velocity**-has become almost mythical in labs. For decades, it has been treated as the shorthand answer to the question, "Is my hood safe?"

But the truth is, that number alone tells you almost nothing. It is like asking the average speed of all the cars on an interstate and assuming every driver is moving at exactly that pace. In reality, some cars are crawling in the slow lane, others are weaving in and out at breakneck speed. The average might be 65 mph, but the driving experience is anything but uniform. Average face velocity suffers the same flaw. It compresses all the complexity of airflow-eddies, drafts, turbulence-into a single, misleading figure.

The Profile: Reading Between the Numbers

Now, compare that to a **face velocity profile**. Instead of one number, a profile maps out dozens of numbers taken across a grid over the sash opening. Each cell of that grid-typically no larger than 12" x 12"-is tested individually, often for 20 seconds at a time, capturing both the average velocity and the variability in that spot.

The result looks less like a neat summary and more like a topographic map of airflow. Peaks and valleys appear. One square may show 120 fpm, its neighbor only 60 fpm. A few inches can make the difference between containment and escape.

Where the average velocity might read a reassuring 100 fpm, the profile reveals a storm of turbulence hiding in plain sight. That turbulence is the hood's true enemy-it is the invisible hand that carries vapors out into the room.

Why It Matters

Research has shown time and again that face velocity alone does not predict containment. Some hoods with "perfect" averages fail tracer gas testing miserably. Others with modest velocities perform beautifully. The deciding factor is not how fast the air is moving, but how evenly and smoothly it moves.

Think of it as water flowing down a stream. If the current is steady and laminar, a boat glides smoothly. But if the stream is broken into whirlpools and rapids, even if the flow rate is high, the boat is tossed around unpredictably. The same is true of fume hoods: turbulence trumps speed every time.

This is why the face velocity profile is so important. It doesn't just average the stream; it shows where the rapids are. It helps us predict where fumes might escape and why a hood that "looks good on paper" might still be a hazard in practice.

The Myth of the Snapshot
Another problem with average face velocity is that it is just that-a snapshot. Take a handheld probe, hold it for a second, write down the number, and you have your reading. But air is alive. Open a door down the hall, walk past the hood, or raise the sash, and the numbers change instantly.

In fact, one study demonstrated that a 24-hour profile of a single grid point yielded over 86,000 readings, varying by as much as 40 fpm from the average. Imagine trusting a single snapshot to summarize all that volatility.

The profile, by contrast, accepts that variability is the rule, not the exception. By sampling multiple points, it paints a picture of airflow under real-world conditions, not just a one-time measurement on a calm afternoon.

Beyond the Numbers: Containment
Ultimately, neither average velocity nor profile is the final word on hood performance. Containment is. That's why ASHRAE 110 includes tracer gas testing and flow visualization, and why new tools like the Tri-Color laser system are so valuable.
But among the velocity tests, the profile stands head and shoulders above the simple average. It is the difference between glancing at the cover of a novel and actually reading the chapters inside.

The Story the Profile Tells

When I perform a profile, I often watch the numbers dance across my logger as I move grid to grid. A few high readings here, a few low ones there, and suddenly the story begins to emerge. Perhaps the supply diffuser above is creating a downdraft. Maybe equipment inside the hood is blocking airflow. Sometimes I see the fingerprint of a poorly balanced system.

The average face velocity would never reveal these truths. The profile does. It is not just a measurement-it is a conversation with the hood, a way of asking, "Where are you strong? Where are you weak? Where might you fail the person standing here?"

Conclusion

So how does a face velocity profile differ from average face velocity? The average is a comfort blanket-a single, tidy number that gives the illusion of safety. The profile, on the other hand, is a diagnostic lens. It exposes the turbulence, the weak spots, the hidden risks.

If I had to choose one test to perform on a fume hood, it would be the profile. Because safety doesn't live in the average. Safety lives in the details, in the airflow that swirls unseen just inches from a scientist's breathing zone. And only a profile tells that story.

Average Face Velocity (AFV) is simply the mean speed of air entering the hood across the sash opening, expressed in feet per minute (fpm). It tells you how fast air is moving on average—but not how evenly it's moving.

Face Velocity Profile (FVP) maps out the distribution and turbulence of that airflow across multiple grid points on the sash opening. It reveals where the flow is too high, too low, or swirling—the turbulence patterns that can cause loss of containment.

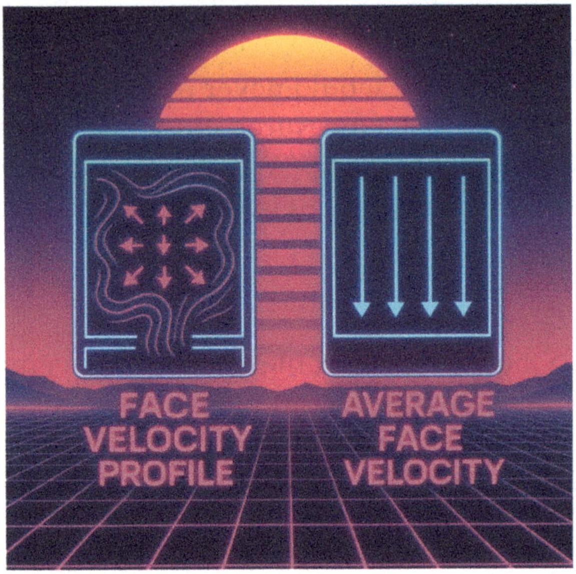

Average Face Velocity = Speed
Face Velocity Profile = Speed + Turbulence + Stability

Chapter 8:
Room Instability

Remember we are taking about a snapshot. So how would handle this?

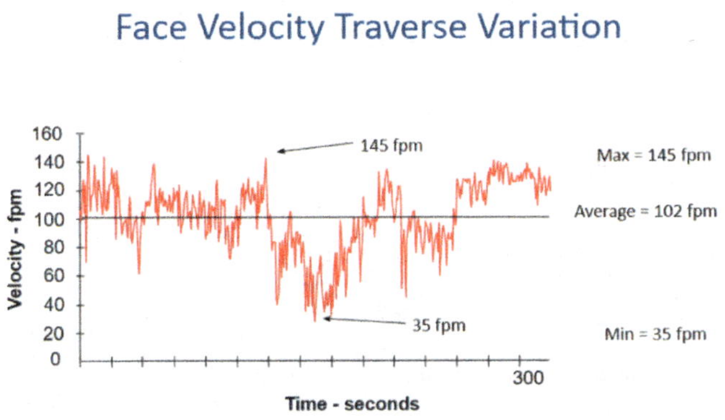

This is a 300 second velocity reading from a single point. It is likely that the hood alarm would be showing a number around 100 fpm. You can see even over this short period that average face velocity was varied. With a high of 145 fpm and low of 35 fpm. If you used the standard of 20 readings over 20 seconds, you would get averages that varied greatly depending on which 20 seconds you average.

This is a real test, not a fabricated chart. And yes, I have seen worse. Lately I have started doing a 300 second face velocity reading while doing the SF6 tests. I am starting to see a pattern. When there is loss of containment, there is usually a shift in face velocity.

In the last chapter we talked about the Face Velocity Profile. While it is mainly looking at turbulence, it is giving us a preview of differential pressure shifts. If the exhaust is not stable that will get telegraphed into the face velocity profile. If there is instability in the supply it will telegraph into the face velocity profile.

When troubleshooting, we have a process to identify the source of the instability. And generally, the answer is never simple. Remember that fans are motor driven. Their speed can vary because of fluctuations in the electrical voltage, loads, and the controls. It is hard to find a room that is really stable.

Even AM test rooms have issues with stability. Lab buildings are much more complex with many more components that can impact stability.

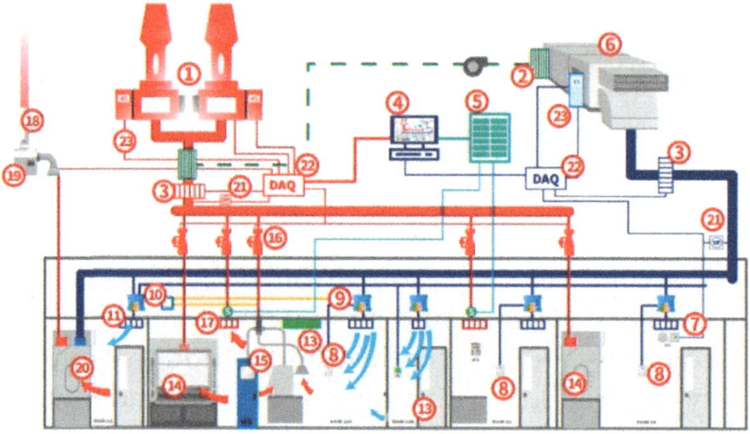

In this diagram you can see that to support 3 hoods there are over 200 components in the Laboratory Ventilation System and any of them can impact the fume hoods performance.
This is why dynamic containment testing is the only way to determine how a fume hood is performing under real world conditions. And even that cannot fully account for the instability in the room and how the user is interacting with the hood.

Chapter 9:
Smoke Visualization - The Oldest Test, Still the Trickiest

If you step back into the laboratory testing rooms of the 1940s, before tracer gas analysis and laser visualizers, there was really only one way to "see" air movement in a fume hood: smoke. It was simple, it was intuitive, and it gave both scientists and engineers the comfort of a visual confirmation. If the smoke appeared to stay in the hood, it was assumed the hood was safe. But like so many things in laboratory ventilation, the reality was more complicated.

The problem with smoke, as it was first used, was obvious to anyone who had ever tried to paint a white picture on a white canvas-you can't see much. Smoke is almost always white or gray, and most hood interiors have white liners. The result is "white on white." The very medium designed to show airflow patterns often blended into the background. Subtle currents, swirls, or reflux of air could be happening right in front of your eyes, yet remain invisible.

Worse yet, traditional smoke sticks or bombs eject smoke at high temperature and velocity. That hot, fast-moving smoke is not a neutral tracer-it's an active disruptor. Instead of simply following the natural airflow inside the hood, it creates its own micro-drafts, pushing and tumbling in ways that would never occur with a real chemical vapor. The tool meant to reveal truth could easily create illusions.

The ASHRAE 110 Take on Smoke
ASHRAE 110-the standard that formalized hood testing-acknowledges smoke's limitations. The protocol calls for two flavors of smoke visualization:

- **Local (small volume) smoke release** – a wand or puff of smoke to probe specific locations, like under the airfoil or around a piece of equipment. This is really seeking to identify turbulence.

- **Large volume smoke release** – flooding the hood interior to observe the general containment and airflow stability.

Both can reveal useful information, but both require judgment. Unlike a face velocity meter or a tracer gas analyzer that spits out a number, smoke testing is qualitative. What one tester interprets as "good containment," another might see as marginal. This is why many experienced testers say smoke visualization is "more art than science." It's about learning to release smoke without disturbing the flow, and about training your eyes to catch the faint wisps that indicate reverse flow. That skill only comes with years of practice.

The Subjective Nature of Smoke
Because smoke testing relies so heavily on interpretation, it is by far the most subjective and least reproducible of the ASHRAE 110 tests. Two testers could perform the same smoke release on the same hood and report different outcomes. Lighting in the room, the color of the sash or liner, even the tester's eyesight all play a role. For that reason, smoke visualization alone has never been considered a pass/fail indicator of safety. At best, it is a diagnostic tool.

Yet despite its shortcomings, smoke visualization continues to be used worldwide. Why? Because it communicates something numbers cannot. If a hood user watches smoke curl out of the sash and into their breathing zone, no explanation is needed. The message is visceral and immediate: "This hood is leaking."

Beyond Smoke - Lasers and Haze

In recent years, innovators have looked for ways to overcome the limits of white-on-white smoke. The most promising solution has been laser-enhanced haze visualization. Instead of dense smoke, a fine, neutrally buoyant haze is released into the hood. Then, planes of colored laser light-red, green, blue-slice through the haze. Suddenly, invisible air currents are painted vividly in three dimensions. A swirl at the sash is no longer a ghostly wisp-it's a bright green plume escaping toward the user's face.

The "Tri-Color Visualization" technique transforms a subjective art into a powerful teaching and diagnostic tool. Laboratory managers can show new users how a careless arm movement pulls contaminants into the room. Engineers can document exactly where containment is lost during a walk-by disturbance. Even experienced testers are often surprised when the lasers reveal leakage that conventional smoke wands never caught.

The Value-and the Limits

So where does this leave smoke visualization? On one hand, it is the most subjective, least quantitative, and least reliable of the ASHRAE 110 toolbox. On the other hand, it is often the most powerful in persuading people to take hood safety seriously. Numbers in a report may be ignored, but once you see smoke-or haze and lasers-escape toward your face, you never forget it.

In that sense, smoke visualization has a unique role. It may not give us definitive metrics of containment, but it does something just as important: it makes the invisible visible. And in the world of laboratory safety, that act of revelation can change behavior, improve vigilance, and ultimately save lives.

We are trying to change the preferred term from Fume Hood Safety to Accident Prevention. This at least makes some for the hazards and risks more visible and a more actionable plan can be developed to manage these risks.

Chapter 10:
Containment Testing
The Test That Really Matters

If we strip away all the noise about face velocity numbers, grids, and smoke wands, what are we really trying to measure? At its core, the entire purpose of a fume hood is simple: **to contain hazards and keep them out of the user's breathing zone.** That is why containment testing has always been, and will always remain, the ultimate yardstick of safety.

Since 1985, ASHRAE 110 has prescribed the use of **sulfur hexafluoride (SF_6)** as the tracer gas of choice. SF_6 is an unusual gas, used mainly as an electrical insulator in high-voltage switchgear. For our purposes, it was adopted because of its unique stability, detectability, and inertness. Paired with the **Q200 leak detector**-a tool originally developed for the power utility industry-testers can detect leaks down to an astonishing **0.01 ppm.** That sensitivity gave us confidence: if a hood passed this test, it was truly containing.

And for decades, it worked. SF_6 became the common language of hood testing across North America, Europe, and Asia. If you wanted to compare hoods, validate performance, or troubleshoot a failure, this was the tool you reached for.
But there was always a catch: SF_6 is one of the most potent greenhouse gases known. Molecule for molecule, it is over **23,000 times more impactful than CO_2** in terms of global warming potential. While the amount used in laboratory hood testing is minuscule compared to power utilities or semiconductor fabs, the optics have never been good. For nearly 20 years, committees, researchers, and manufacturers have searched for alternatives.

So far, **no substitute has measured up.** Propylene and isopropyl alcohol (IPA) vapors have been proposed. Both are detectable. Both are more environmentally acceptable. But they are also **flammable,** which introduces an entirely new dimension of risk. Testing labs would have to redesign ejectors, retrain staff, and mitigate fire hazards-all to adopt a gas that still doesn't behave exactly like SF_6.

This brings us to the thorny question of **equivalency.** Imagine this: yesterday your hood "passed" containment with SF_6. Today, you test the same hood with propylene, and it "fails." Which one do you trust? Did the hood suddenly become unsafe, or are we simply moving the goalposts? As one ASHRAE committee member put it, **"If you change the gas, you change the test-and if you change the test, you change the results."**

That is why some of us advocate a **dual-gas approach.** For a transition period, hoods could be tested with SF_6 and propylene side-by-side, building a correlation database. Over time, once confidence is established, SF_6 could be retired. Until then, SF_6 remains the only tracer with decades of validation behind it.
Meanwhile, the industry has not stood still. **Tri-Color laser visualization** is emerging as a powerful complement to tracer gas. Using haze and multi-color laser light curtains, Tri-Color makes the invisible visible. For the first time, users can actually see containment being lost: wisps of air curling out of the sash opening, or vortices pulling vapors back toward the breathing zone. The impact is visceral. Researchers who once relied only on ppm numbers now have a real-time picture of airflow failures.

The promise of combining methods is compelling:

- **Use Tri-Color as a first screen.** If no loss of containment is visible, skip tracer gas.
- **If Tri-Color shows leaks,** confirm with SF_6 (or its eventual replacement).
- **Use data together,** giving a richer, more nuanced picture of performance.

The reality is that hoods fail containment far more often than most organizations want to admit. Studies suggest that **nearly half of all hoods tested in the field show some degree of containment loss**. Causes vary: poor design (25%), bad room conditions (50%), or simply user work practices (25%). But the lesson is clear: containment cannot be assumed. It must be tested.

The standards reflect this. Both **ASHRAE 110** and **EN 14175** are built around validating containment as the ultimate measure of safety. Z9.5 echoes the same principle: exposure control devices exist for one reason-to prevent harmful airborne exposures. OSHA requires employers to ensure hoods are "functioning properly", which in plain English means: they must contain.

So where does that leave us today?
SF_6 is still the gold standard, despite its environmental baggage.

- **Propylene and IPA are on the horizon,** but questions of safety and equivalency remain unresolved.
- **Tri-Color laser visualization is bridging the gap,** giving us a tool that is safer, more accessible, and educationally powerful.

My own belief is that the future is hybrid. We should keep SF_6 in the toolbox for R&D, troubleshooting, and correlation studies. But for routine annual testing, Tri-Color combined with smart velocity profiling can give us a safer, more sustainable path forward.

Containment testing is not just another checkbox in a certification report. It is the moment of truth. The question it answers-is this hood really protecting the user? is the only question that matters. And until we can answer it with clarity and confidence, all the numbers in the world won't keep a chemist safe.

Fog machines generate high velocity hot smoke that can easily disrupt the natural airflow. So what you are seeing may be more the effects of the smoke generation than the hood.

Chapter 11:
VAV Response Time

When Variable Air Volume (VAV) systems entered the laboratory world in the 1980s, they seemed like the perfect solution. Why waste energy pulling a constant volume of air through a hood when the sash might be closed half the time? VAV systems promised smarter, more efficient labs-adjusting airflow as demand changed. But like most "smart" systems, they introduced new challenges.

One of those challenges is something most users don't even know exists: **VAV response time**. It's hidden in the background, quietly dictating whether a fume hood contains or leaks during those critical few seconds when room and hood conditions shift.

A Bucket of Water and Two Valves

Let's step back and imagine a bucket of water. The bucket has a mark halfway up - our "waterline." Now imagine two valves: one at the bottom to drain water out (exhaust) and one at the top to refill it (supply). The goal is simple: keep the water exactly at the line.

Sounds easy, right? Until you start turning those valves. Open the bottom valve too quickly and the level plunges. React too slowly with the top valve, and the bucket runs low. Overshoot with the supply, and you're sloshing over the mark.

This dance of inflow and outflow is exactly what's happening in a VAV-controlled lab. The "bucket" is the laboratory, the waterline is the pressure balance, the drain valve is the exhaust, and the top valve is the supply. When the sash opens, the demand for exhaust shoots up. The system has to sense it, react, and rebalance-before containment is lost.

Why Response Time Matters

Containment is driven by **differential pressure**-that subtle pull keeping contaminants inside the hood. VAV systems work by opening and closing valves based on sash movement, door openings, and room activity. But nothing happens instantly. When a sash is raised, the system sees a surge in open area, meaning more air must be drawn through the hood face. The exhaust valve begins to open, and the supply must match it. But in the few seconds it takes for those dampers to move, the balance is unstable. Pressure may drop. Cross-drafts may overwhelm the hood. Vapors that were once safely inside can spill out.

ASHRAE 110 recognized this decades ago, adding **VAV response time testing** into its protocol. The idea was simple but powerful: measure how quickly a hood and its controls recover after a sash movement. The longer the lag, the greater the risk of exposure.

What the Numbers Tell Us

Field experience has shown us that containment failures are rarely about "average" airflow. They're about the dynamics-the turbulence, the delays, the little windows of instability when the hood is vulnerable.

Research has revealed that even a short delay in response can have a measurable effect on containment. During that lag, the **face velocity drops**, the **differential pressure weakens**, and the "air curtain" across the sash opening falters. Users leaning into the hood may never notice. But an SF_6 detector, or a Tri-Color laser test, will expose exactly what's happening.

Real-World Examples
I once tested a hood in a brand-new facility that had invested millions in state-of-the-art VAV controls. On paper, everything was perfect. The system balanced beautifully under steady conditions. But when I raised the sash quickly, I watched the Tri-Color haze spill straight out into the room. The response time was nearly four seconds-an eternity in containment terms. Four seconds may not sound like much, but in a lab, it can be the difference between exposure and protection. Imagine working with a volatile solvent or toxic gas, opening the sash to adjust equipment, and unknowingly stepping straight into a plume of contaminated air.

Standards and Best Practice
Both **ASHRAE 110** and **SEFA's guidance on Exposure Control Devices** emphasize the importance of dynamic testing, particularly for VAV systems. The standards acknowledge that static tests aren't enough-labs live in motion. Sashes rise and fall, doors slam, people walk past. If a system can't keep up, the hood can't be trusted.

Modern test protocols, including the **Tri-Color dynamic containment protocol**, go further by simulating real-world challenges: rapid sash movement, door openings, and even walk-bys. These dynamic elements expose weaknesses that steady-state numbers hide.

What This Means for You
If you're an EHS officer, a lab manager, or simply a hood user, here's the takeaway: **don't assume your VAV hood is safe just because it's "smart."** Ask whether VAV response time has been tested. Insist on dynamic tests, not just static airflow readings.

Think of it this way: in aviation, we don't just test whether an autopilot works on a straight, level flight. We test how it responds to turbulence, sudden climbs, and emergency maneuvers. A fume hood deserves no less.

The Bigger Picture

The story of VAV response time is really the story of laboratory safety itself. On paper, everything can look perfect. But the real world is messy-people move, doors swing, equipment changes, and air is never still.

By paying attention to response time, we acknowledge that safety isn't about averages or snapshots. It's about resilience-the ability of a system to adapt quickly and keep protecting even when conditions shift.

That's why VAV response time isn't just a technical detail buried in ASHRAE 110. It's a critical measure of whether your hood is truly doing its job: keeping the invisible hazards where they belong-inside the hood, not in your lungs.

A point for EHS professionals: We say we are testing the hood, but in reality, the hood hasn't changed, it is the interface with the room conditions that we are testing. Given most modern buildings are VAV, we are looking at VAV response time and how it impacts fume hood performance. If the hood had a good AI test, but now has lesser performance, then the VAV may need to be recalibrated and the room rebalanced.

Chapter 12:
Cross Drafts - The Invisible Saboteur

If you've ever stood on a windy street corner, you know that air doesn't always move in a straight line. It swirls, gusts, and slips around obstacles. The same thing happens inside laboratories, except the stakes are far higher. The invisible movement of air can be the difference between a chemical being safely captured inside a fume hood-or escaping into a researcher's breathing zone.

Air is lazy. It always looks for the path of least resistance. And when it finds a shortcut across the face of a hood, it can sabotage containment. We call these disruptive flows cross drafts.

The Subtle Enemy

Imagine you're carefully working at a hood, trusting that the gentle pull of 100 feet per minute is keeping you safe. Overhead, a supply diffuser kicks on, sending a stream of cool air down at 1,000 to 2,000 fpm. That jet doesn't care about your experiment or your safety-it slams into the sash opening like a river crashing through a dam. In an instant, the orderly draw of air into the hood is broken apart, turbulence erupts, and vapors that should have been contained are shoved back into the room.

This is the unseen reality of many laboratories. Most supply registers are overhead, delivering air at steep angles. If the wrong register is selected, or if the diffuser isn't balanced properly, you've got a recipe for containment failure before a single test tube is even lifted.

Why Cross Drafts Matter

ASHRAE 110 acknowledged this decades ago. The whole point of the "As Used" (AU) tests is to see how hoods behave in real rooms, with doors opening, people walking by, and supply systems humming. Studies have confirmed what field testers know by experience: even modest cross drafts-just 30% of the hood's average face velocity-can cause containment failures.

Think about that: if your hood is set to 100 fpm, a sideways current of just 30 fpm is enough to disrupt containment. Now remember that diffusers regularly pump air out at ten times that speed. It doesn't take much misdirected supply air to compromise safety.

Seeing the Problem

One of the challenges is that cross drafts are invisible. A researcher can stand at a hood, see the sash down, hear the whoosh of air, and feel confident. Yet a supply diffuser may be quietly pushing their safety margin over the edge.

That's why visualization tools have been so transformative. With smoke wands-or better yet, laser-enhanced Tri-Color systems-you can literally see how a cross draft tears across the sash and pulls contaminants out. Suddenly, what was once a hidden airflow dynamic becomes a visible, undeniable risk. It's not a theory; it's right there in front of your eyes.

Real-World Consequences

Cross drafts are not just an academic problem. Walk into almost any lab and you'll find them in action:

- A researcher working near a lab door feels a rush of air every time someone in. This change in room pressure can cause loss of containment.
- Two fume hoods side by side, with one sash moving up and down rapidly, create currents that buffet the other hood's opening.

- In older labs, where overhead air was designed for comfort cooling, registers blow directly at the hood face, practically daring containment to fail.

Every one of these scenarios has been linked to containment losses in real tests. In fact, about half of fume hood failures can be traced back not to the hood itself, but to the room conditions that surround it.

Designing Against the Draft
The good news is, cross drafts can be managed. Proper diffuser selection, placement, and balancing are critical. Low-induction diffusers that gently wash air across ceilings instead of blasting it downward are far better for laboratory safety. Careful room pressurization and door placement also help.

Standards like ANSI/ASSP Z9.5 remind us that it's not enough to specify a hood-you must consider the entire laboratory ventilation system. A great hood in a bad room is like a racecar on a gravel road: performance doesn't matter if the conditions undermine it.

The Tester's Perspective
As a tester, identifying cross drafts is one of the most important parts of evaluating hood safety. It's not just about putting an anemometer in the sash and logging numbers. It's about watching how the room interacts with the hood. It's about asking: What's the real airflow environment here?

Cross drafts are the silent saboteur of fume hood performance. You don't see them on a spec sheet. You don't notice them until something goes wrong. But if you know where to look-and if you test wisely-you can catch them before they cause harm.

Chapter 13:
Building on ASHRAE 110 with Tri-Color

When ASHRAE 110 was written, its authors understood something important: a fume hood sitting quietly in a perfect test room isn't the same as a hood in a bustling laboratory. People walk past, doors swing open, nearby hoods are used, and sashes move up and down. The standard itself even says:

"It is important to evaluate the performance of the laboratory hood under dynamic conditions. Specific operations, such as a pedestrian walking past the hood, laboratory doors opening, and specific actions at the hood, are only a few of the challenges that could be expected at the hood. This test method addresses only the dynamic challenge of sash movement."

Some read that as a suggestion. I took it as a mandate.

The Early Days of Lasers

Back in 2007, I started experimenting with lasers. The idea was simple: if air movement is invisible, what if I could make it visible? The execution, however, was anything but simple. Hundreds of tests later, I had learned to read the faintest flicker of airflow and see containment breaches that no smoke test could reveal. The laser was powerful-but it demanded expertise. The untrained eye could look at the same scene and miss the story the air was telling.

For years, this remained my secret weapon: a tool to troubleshoot hoods, diagnose failures, and teach those who wanted to see what air was doing. But it wasn't practical for everyday users.

The Pandemic Pause
Then came 2020. Like everyone, I was grounded, working more from home, reflecting on conversations I'd had for years. One comment stuck with me: **"Testing is too complicated and expensive."** That's why many labs weren't doing it.
That thought nagged at me. Could I simplify what I was seeing with lasers-simplify it enough that any lab manager, any EHS officer, even a hood user, could perform the test and instantly understand the results?

From Complexity to Clarity
After a year of development, the answer emerged: **Tri-Color**. Instead of one laser, we used six, in three different colors, forming curtains of light. Add a harmless visualization haze, and suddenly airflow patterns leap into view. Red illuminates the internal currents of the hood. Green shows what happens at the sash boundary-the user's breathing zone. Blue highlights the impact of movement or turbulence. When vapors escape, you see it immediately. No interpretation required. What once took an expert's eye became obvious to anyone standing in front of the hood.

Building on ASHRAE 110
ASHRAE 110 gave us three pillars:

1. **Face velocity profile** - a grid of measurements across the sash.
2. **Flow visualization** - using smoke to see patterns.
3. **Tracer gas containment** - the gold standard, but expensive, complex, and reliant on SF_6, a greenhouse gas.

Tri-Color doesn't replace these. It builds on them. We still do s face velocity profile. We still use local smoke. But then we add dynamic challenges-the very ones ASHRAE mentioned but never fully incorporated:

- A pedestrian walk-by
- A nearby door opening
- Rapid sash movement
- A hand reaching in and out
- Materials being moved from the hood

These are the moments when hoods fail in real life. With Tri-Color, failures aren't numbers on a report-they're streaks of light and haze escaping into the lab, plain for everyone to see.

The Educational Effect

One of the unexpected benefits of Tri-Color is how it changes behavior. Traditionally, half of fume hood containment failures trace back not to the hood itself, but to room conditions or user practices. Teaching this has always been a challenge.
But when a chemist sees their arm motion drag air out of the hood, they don't need a lecture. They adjust. The lesson sticks.
As I often say: "If you can see a problem, you can fix it."
Tri-Color is as much an educational tool as a testing protocol.

Toward a De Facto Standard

ASHRAE 110 isn't going away-it remains the foundation. But just as Z9.5 broadened the scope to the entire ventilation system, Tri-Color extends ASHRAE 110 into the real world. Already, more and more organizations are adopting Tri-Color protocols, not just for certification but for training, commissioning, and ongoing monitoring. In many labs, it has become a de facto standard because it answers the question most people really care about: **Is my hood safe right now, in the way I actually use it?**

ASHRAE 110 gave us the foundation, but it left room for evolution. Tri-Color is that evolution-making the invisible visible, the complex simple, and the abstract personal. It doesn't replace the standard; it fulfills its spirit.

For decades we've tested hoods in controlled ways and handed users a report they barely understood. With Tri-Color, the hood itself tells the story. And once you've seen air move in living color, you'll never look at a fume hood the same way again. Tri-Color is more actionable than a printed report with numbers.

Chapter 14:
New Frontiers in Fume Hood Testing: Lasers, Colors, and Greener Alternatives

For decades, fume hood testing lived in a fairly narrow world: measure some face velocity, release a puff of smoke, run an SF_6 tracer gas test, and sign off. But the 21st century has forced us to think differently. We've begun asking deeper questions: Does face velocity really tell us containment? Can we rely on a greenhouse gas to measure safety? Is there a way to see the invisible?

That's how a new frontier was born-one that looks less like yesterday's engineering manual and more like science fiction. Lasers, color-coded airflow patterns, and a push toward greener tracer methods are reshaping how we evaluate safety inside the hood.

The Trouble with the Old Ways
ASHRAE 110–2016 remains the benchmark, but even its own foreword admits limits: the test method only directly addresses sash movement as a dynamic challenge. Real labs are far more chaotic-doors swing open, people walk past, equipment clutters the work zone, cross-drafts stir the air. Relying on a single static tracer gas test was like judging a marathon runner by their posture at the starting line.

And then there was SF_6. For nearly forty years it has been the "gold standard" tracer gas. Its detectability is unrivaled, but its environmental profile is grim. With a global warming potential over 20,000 times that of CO_2, California and parts of Europe have already restricted its use. We were left with a paradox: the best way to measure safety was also a climate liability.

Enter the Lasers
In 2007, laser visualization started to sneak into my toolbox. It began as an experiment-how do you make the invisible visible without disturbing airflow like hot smoke does? By creating a fine haze and shining light sheets across the hood, airflow suddenly became a dance of streams and eddies. It took patience to interpret, but for the first time testers could see containment.

Fast forward to today, and that experimental trick has grown into **Tri-Color Laser Testing**. Imagine a hood bathed in red beams tracing internal flows, green curtains standing guard at the sash, and blue light revealing details of the vortex. When the green lasers light up outside the sash, you know instantly: vapors are escaping. No complex charts required. Even the most novice lab user understands what they're seeing.

The effect is dramatic. Researchers, safety officers, and even skeptical faculty stand in front of a Tri-Color test and immediately grasp why their hood matters. The colors don't just diagnose-they educate.

Dynamic Testing Becomes Real
Tri-Color didn't stop at pretty pictures. It took ASHRAE 110's hint about "dynamic challenges" and ran with it. Walk-bys, door openings, fast sash movements, even quick hand motions inside the hood are now part of the script. And when you watch the colored light bend, swirl, or spill outward during those actions, you understand in your gut why behavior matters.
In truth, half of all containment failures are not due to hood design at all-they're due to room conditions or user practices. Dynamic visualization makes that painfully obvious. For the first time, we can train users not by lecture, but by showing them exactly what their actions cause in real time.

Searching for Greener Alternatives
But what about tracer gas? That question is still unsettled. Replace SF_6 and you're not just swapping chemicals-you're changing the very backbone of the test. Efforts with CO_2, nitrous oxide, and other candidates continue, but none yet offer the same sensitivity without significant drawbacks. Some argue we should reserve SF_6 for factory R&D and troubleshooting only, while moving routine field tests toward visualization and velocity profiles. Standards like Z9.5 have already encouraged more flexible, holistic approaches.

It's a transition moment: the industry knows it cannot lean forever on a gas that regulators want gone, but it has not yet landed on a perfect replacement.

A Hybrid Future
So where does that leave us? In practice, a hybrid model is emerging:

- **Face Velocity Profiles** still provide a map of turbulence-but not the verdict on safety.
- **Laser Visualization** brings clarity and training value that numbers alone can't.
- **Tracer Gas (SF_6 or alternatives)** remains a forensic tool when precision is needed.
- **Greener Testing Protocols**, like Tri-Color, fill the gap for routine verification without environmental guilt.

Think of it as an evolution: from measuring to seeing, from single-number pass/fail to multi-layered understanding.

Why It Matters

At first glance, lasers and colors may look like theatrics. But don't mistake spectacle for show. What's happening is a profound shift in how we think about safety. Instead of hiding risk in tables and reports, we bring it into the open where everyone can see it. Instead of relying on a single metric, we embrace a spectrum of indicators-dynamic, visual, environmental.

The result is not just better data, but better behavior. A researcher who has seen a plume escape when they leaned too close is a researcher who will remember. And that is where real safety begins.

Closing Thoughts

The future of fume hood testing will not look like the past. It will be brighter-literally, with beams of red, green, and blue cutting across the lab. It will be greener-stepping away from gases with catastrophic warming footprints. And it will be more human-designed not just for engineers with instruments, but for everyday lab users who deserve to know, in the simplest terms possible, whether their hood is protecting them.

Because in the end, this frontier is not about lasers or gases. It's about seeing what was once invisible, and in doing so, exposing hazards that keep people safe.

Chapter 15:
The Human Element
Testing and Training Go Hand in Hand

When people talk about laboratory safety, the focus is usually on the equipment. The fume hood. The ventilation system. The airflow monitors. These are critical pieces, but we often forget one truth that can't be engineered away: people are at the center of the system.

A hood may be perfectly designed and tuned, but its real-world safety depends on how it is used. Research shows that **about 25% of containment failures are caused by user setup and work practices**. Another 50% trace back to room conditions and ventilation system interactions, many of which can be influenced-positively or negatively-by user behavior. That's the "human element" at work.

Testing as Training

For decades, testing was seen as a purely technical exercise-something a specialist did with instruments, probes, and reports full of numbers. But testing can also be one of the best forms of training. When users see how their own actions impact containment, the lessons stick.

This is where modern visualization tools like **Tri-Color laser testing** have changed the game. Instead of an invisible concept, airflow becomes visible. A researcher leans into the hood too far, or places equipment in a way that blocks the baffles, and the colored laser haze reveals how fumes swirl out into the room. Suddenly, what was once theoretical becomes unforgettable.

As one lab manager put it after a Tri-Color demonstration: "It wasn't the numbers on the report that changed behavior. It was seeing the red and green lasers light up where they shouldn't. That's what got people's attention."

The OSHA Link

Regulators recognize the same reality. **OSHA's laboratory standard requires employers not just to provide fume hoods, but also to train employees on their proper use**. A hood that passes ASHRAE 110 tests still won't protect if the sash is left wide open, if rapid arm movements stir up turbulence, or if chemical containers are stored inside like it's a closet.

That's why testing and training belong together. A test without training is a missed opportunity. And training without testing is just theory, untethered from the real performance of the hood in its actual lab environment.

Lessons That Stick

I remember walking into a teaching lab where the students had been given a short demonstration: a simple smoke test at the hood face. At first, they joked, "Oh, the hood sucks-it'll all stay in." Then the instructor cracked the door open, and the smoke suddenly billowed across the room. The laughter stopped. The students leaned in. The message was received in a way no lecture alone could deliver.

This is the heart of the human element: **we learn best when we see cause and effect.** If we don't understand the hazards, we can prevent accidents.

Building a Culture of Awareness

Testing and training together create something bigger than compliance: they build culture. A culture where users feel responsible not just for their experiment, but for their safety and the safety of their colleagues.

- **Testing validates** that the equipment works.
- **Training teaches** the behaviors that keep it working.
- **Together, they empower users** to spot problems, report issues, and make corrections before accidents happen.

When a hood is labeled with a QR code that links to its test history and a short safety video, the line between testing and training disappears. Each time a user scans that code, they're reminded that safety isn't static-it's a shared responsibility, verified in real time.

The Bottom Line

The most sophisticated standards-ASHRAE 110, Z9.5, SEFA 1 outline test protocols and performance targets. But they all converge on the same truth: laboratory safety doesn't end with numbers on a report. It lives in the hands, eyes, and decisions of the people who stand in front of the hood every day.

Testing confirms the hood is capable of safety. Training ensures that safety is actually achieved. One without the other is incomplete. Together, they are the foundation of a safer lab.

In reality, the testing process should be a teaching and training experience. I tell the testers I certify that 50% of their role is education.

Fume Hood Testing and User Training go hand in hand

Chapter 16:
Making Labs Safer
One Hood at a Time

When you walk into a laboratory, the most dangerous hazards are rarely the ones you can see. Bottles of acids, flammable solvents, and powders may sit on a shelf, but the true threat often lies in the invisible vapors and airborne contaminants that are released the moment those containers are opened. For more than half a century, the chemical fume hood has been our primary line of defense against those unseen enemies. But a hood, by itself, is not enough. It only becomes a true guardian when it works in harmony with the ventilation system that supports it and with the person standing in front of it.

That's the lesson we've learned-sometimes the hard way-through countless studies, tests, and even tragedies. About half of the hoods tested in field surveys fail at least one containment evaluation. And the failures aren't usually because the hood was poorly designed. In fact, only about a quarter of failures can be traced back to the hood itself. Half stem from the surrounding room and ventilation system, and the other quarter from user work practices. In other words, safety depends on the whole ecosystem: the hood, the system, and the user.

Seeing What Was Once Invisible

For decades, laboratories relied on face velocity-measuring how fast air moves into the hood-as the gold standard for safety. But we now know that velocity alone doesn't guarantee containment. Air is like water: it doesn't just flow, it swirls, eddies, and churns. A hood with "perfect" velocity can still spill vapors if the room is turbulent or if a researcher blocks the airflow with their body. Much like a rip tide in the ocean.

This is where new visualization tools, like Tri-Color laser testing, have changed the conversation. For the first time, scientists and safety officers can actually see containment-or the loss of it-in real time. Green laser curtains across the sash will light up if vapors leak into the room. Suddenly, what was once an abstract number on a meter becomes a vivid, unforgettable image. Researchers can see how a quick hand movement, an open lab door, or a colleague walking behind them can pull vapors right into their breathing zone.

The User in the Equation

A fume hood is not a magic box. It is a tool-one that demands respect and training. OSHA reminds us that hood operators are required to take measures to ensure their hoods are functioning properly. Yet in too many labs, hoods are treated like background furniture. People pile equipment inside, block the airflow with equipment and chemicals, or work with the sash wide open because it feels more convenient. These are not just bad habits-they are invitations for accidents.

That's why training is just as vital as testing. A Tri-Color demo can be worth more than a hundred pages of safety manuals. When a researcher sees vapors swirl out of the hood because of the way they set up their experiment, they never forget it. And when they change their behavior-lowering the sash, working six inches inside the hood, avoiding clutter-the difference is immediate and visible.

The Bigger Picture

Standards like ASHRAE 110, ANSI Z9.5, and SEFA guidelines remind us that fume hoods are not stand-alone devices. They are part of a laboratory's exposure control strategy. Just as a seatbelt is one element of car safety, the fume hood is one layer of the hierarchy of controls. It works best when combined with well-designed ventilation, good laboratory layout, clear procedures, and informed users.

When we test a hood, we aren't just checking a box for compliance-we're validating that the entire system is protecting the people who trust it with their lives. And when we train users, we are empowering them to become active participants in their own safety rather than passive bystanders.

One Hood at a Time

We may never get to a world where every lab is perfectly safe, but progress comes hood by hood, lab by lab. Each time a hood is properly installed, tested, and its users trained, a laboratory becomes a safer place. Each time a myth is dispelled or a researcher sees containment loss for the first time, awareness grows.

Making labs safer isn't about sweeping reform or flashy new gadgets. It's about steady, consistent improvements-small victories that add up over time. Continued improvement. It's about recognizing that safety isn't a product, it's a practice. And it's about committing to the idea that even if we can't fix every lab in the world, we can make this hood safer, this lab safer, and in doing so, protect the people who will unlock the next great discoveries.

That's how change happens-one hood at a time.

Chapter 17:
Containment Self-Validation
The Magic Button

Imagine if every fume hood came with a "magic button." You're in the lab, working with something volatile, and suddenly you wonder-is my hood really protecting me right now? Instead of calling in a tester, waiting for a certification cycle, or second-guessing your own setup, you could simply press that button and instantly see if containment was intact.

That sounds futuristic, doesn't it? But here's the surprise-it's not the future. It's here today.

How the Hood Becomes Its Own Tester
Modern innovations have given us hoods that carry their own testing equipment. Built into the hood itself is a haze generator-think of it as a miniature, built-in smoke machine-that can release a harmless mist inside the fume chamber at the touch of a button. This haze is picked up by the airflow just like vapors from a chemical experiment, following the same airflow paths. Mounted right in the sash track are lasers that create a bright light curtain. Together, haze and laser reveal the invisible. If air or vapor slips past the containment barrier, it crosses that green light curtain and glows, making loss of containment visible to the user in real time. Suddenly, the fume hood transforms from a silent partner you hope is working into an active guardian you can check at will.

A New Kind of Peace of Mind

Think about what this means in the lab.

You catch a faint odor and wonder if something's escaping. Instead of ignoring it, you hit the button. In a few seconds, haze swirls across your setup, and the lasers show whether the hood is keeping everything inside-or whether your arrangement has compromised containment. Maybe the bottles you stacked too close to the sash created turbulence. Maybe cross-drafts are tugging at the air. Whatever the issue, now you can see it. And once you can see it, you can fix it.

This is self-validation: the hood itself letting you know if it's protecting you.

From Static Testing to Dynamic Reality

Traditional testing methods-face velocity readings, smoke wands, and tracer gas challenges-remain important, but they're snapshots in time. They tell you how the hood performed when tested by someone else. Containment self-validation changes the game by letting users check performance whenever they want, under actual as-used conditions.

That's vital because studies have shown that nearly half of hoods fail containment testing in practice-not because of the hood design, but because of room conditions or user behavior. A self-validating hood bridges that gap between annual certification and daily use.

Training, Awareness, and Empowerment

There's another benefit: education. When users push that button and see their setup causing loss of containment, they learn instantly how their behavior affects safety. It's a powerful teaching tool. Rather than abstract warnings in a safety manual, the lesson becomes visual, immediate, and personal. Researchers gain situational awareness that lasts far longer than a lecture.

Toward Smarter, Safer Labs

In many ways, containment self-validation is part of the evolution toward "smart hoods." Just as cars now alert drivers when they drift out of a lane, fume hoods are learning to alert users when containment is compromised. The goal is the same: reduce accidents, protect lives, and turn unseen hazards into visible warnings.

The magic button isn't just a clever gadget-it's a glimpse into the future of lab safety. A future where users don't have to hope their hood is working. They can know.

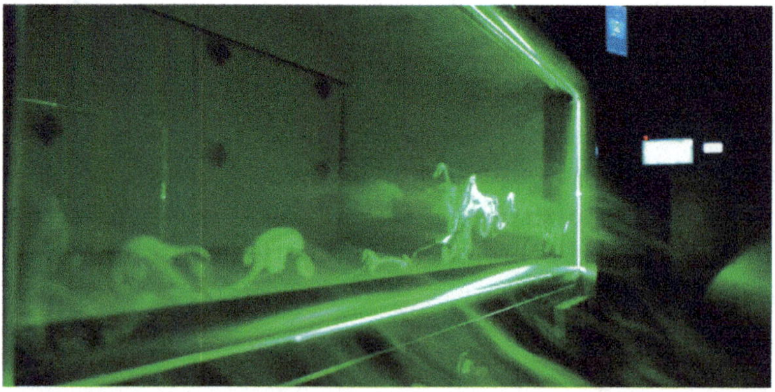

It is like have Tri-Color built-in and ready to test in a moment's notice.

Chapter 18:
International Perspectives
EN 14175 and Beyond

Whenever I travel or speak at international conferences, the same question always comes up: "Which standard is better-ASHRAE 110 or EN 14175?" And I always give the same answer: neither.

That may sound evasive, but it isn't. The truth is that ASHRAE 110 and EN 14175 are not two versions of the same exam where one student scores higher than the other. They are more like two different sports-say baseball and soccer. Both are competitive, both demand skill, but each tests different abilities. A champion in one sport doesn't automatically excel in the other.

The United States and Europe approached the fume hood problem from different directions, guided by their own histories, cultures, and regulatory environments. The result? Two standards that reveal different truths about the same piece of equipment.

How We Got Here: The American Path

In the early 1980s, the U.S. was in the midst of a laboratory safety reckoning. OSHA had begun to exert influence in labs. Universities were building large science complexes. And industry was relying on chemistry more than ever.

But there was no agreed-upon way to measure whether a fume hood was actually protecting the user. One university would claim its hoods were safe based on face velocity readings. Another would do a bit of smoke testing. Manufacturers boasted about "superior containment" without any standardized proof.

Out of that confusion came **ASHRAE 110 (1985)**. The committee wanted a **repeatable, precise method**-a test that could be performed in a controlled room and yield data that could be compared across hoods. Their chosen tool was tracer gas.
They selected sulfur hexafluoride (SF_6), a gas that is easy to detect at parts-per-billion levels. The procedure was simple in concept but sophisticated in practice: release the tracer gas at a specified rate, position a detector at the operator's breathing zone, and measure whether any escaped.

The genius of this test was its **sensitivity**. Even the tiniest leak could be captured in a number. Engineers could now tweak sash angles, baffles, and airfoils and see measurable differences. ASHRAE 110 gave the industry a **laboratory-style instrument** for evaluating containment.

But there was a trade-off. Because the test was so controlled, it didn't fully capture what happens when real people use hoods in real laboratories with cross-drafts, slamming doors, or cluttered setups.

How We Got Here: The European Path
Europe's context was different. Many of their labs were older, built into stone and brick facilities that had to accommodate modern science. Energy costs were higher, and regulations more centralized.

When European committees developed **EN 14175 (1990s–2000s)**, they asked a different question: not just "Does the hood contain under ideal conditions?" but "How stable is the hood under stress?"

EN 14175 focuses heavily on **robustness testing**. Instead of just monitoring tracer gas leakage in a quiet, controlled room, the European protocol introduces deliberate disturbances:

- Cross-drafts blowing across the face of the hood
- Rapid sash movements
- Movements simulating an operator walking by

The measurements are not always as fine-grained as ASHRAE 110, but they reveal something equally important: whether the hood can hold containment when the room is turbulent.
If ASHRAE 110 is the **microscope**, EN 14175 is the **wind tunnel**. One tells you about precision, the other about resilience.

Two Philosophies, One Goal
So, which is better? That's like asking whether it's better to measure the width of a crack with a laser micrometer or to shake the wall and see if it crumbles. Both are valuable.

- **ASHRAE 110** asks: Can I detect any leakage at the operator's breathing zone under tightly controlled conditions?
- **EN 14175** asks: Can the hood withstand realistic disturbances and still protect the user?

Each test shines a light on weaknesses the other might miss. A hood that aces ASHRAE 110 might still be vulnerable to cross-drafts. A hood that passes EN 14175 could still leak small amounts of gas that only a sensitive detector would catch.

Other Voices Around the World

The story doesn't stop with the U.S. and Europe. Around the world, other standards have emerged-sometimes inspired by ASHRAE or EN, sometimes blending elements of both.

- **Canada (MD 15128)**: Canada's federal labs created their own protocol, which leans closer to EN in stressing dynamic challenges but is rooted in the North American tradition of detailed documentation.
- **Japan (JIS Standards)**: Japan, with its culture of precision engineering, adopted methods that combine tracer gas sensitivity with rigorous design criteria.
- **China and India**: In recent decades, as labs have exploded in number, both nations have adapted hybrids. Many new Chinese standards mirror EN, but with growing interest in ASHRAE-style tracer gas testing for R&D and compliance in multinational companies.
- **ANSI/ASSP Z9.5 (U.S.)**: While not a fume hood test standard, Z9.5 lays out the holistic management of laboratory ventilation systems. It reinforces that a hood is only as safe as the system around it.
- **SEFA 1 (U.S.)**: The Scientific Equipment and Furniture Association focuses on construction and design guidance, not testing, but it complements ASHRAE 110 in the marketplace.

The result is a global patchwork quilt. Some labs are stitched into ASHRAE, others into EN, and more and more are trying to integrate both.

Case Stories from the Field

I once visited a laboratory in Germany where the hoods had all passed EN 14175 testing. They were robust, with aerodynamic side walls and careful airfoils. But when we brought in a U.S. team with a tracer gas detector, they were surprised to find measurable leakage at the operator's breathing zone under quiet conditions. Nothing dramatic, but still detectable.

On the flip side, I worked with a U.S. university that had invested heavily in hoods that had impeccable ASHRAE 110 numbers. When we walked past the hoods during a smoke test, you could see plumes spill into the lab. The tracer gas hadn't picked it up because the test wasn't designed to simulate that disturbance.

Both labs thought they were "safe" because they passed "the standard." In reality, each was only seeing part of the picture.

Lessons for Today

The real takeaway is not that one standard is superior. The lesson is that **safety is multi-dimensional**.

- Precision matters. Tiny leaks can mean chronic exposures.
- Robustness matters. Real labs are messy, noisy, and unpredictable.
- Context matters. What's "safe enough" in one country may not meet expectations in another.

The best labs recognize this. Many global pharmaceutical companies, for example, test to **both ASHRAE 110 and EN 14175**. They want the microscope and the wind tunnel. And increasingly, international projects are pushing toward a **"dual-language" of testing** where results can be translated across borders.

The Future of International Standards
Where do we go from here? Already, we see pressures reshaping standards:

- **Environmental impact**: The greenhouse concerns around SF_6 are pushing for alternatives. Europe has been quicker to adopt restrictions, which is influencing the conversation globally.
- **Dynamic testing**: More emphasis is being placed on disturbances, not just static conditions.
- **Visualization tools**: Innovations like Tri-Color laser testing are bridging the gap-offering both sensitivity and real-world visualization.

I believe the future lies in **integration**, not competition. The next generation of standards may borrow the best of both-precision tracer gas methods, dynamic disturbance simulations, and powerful visual tools.

The next time someone asks, "Which is better, ASHRAE 110 or EN 14175?" the answer is clear: neither, and both. Each tells a part of the story. Each grew out of its own soil. It is like comparing wines. Even though they might be from the same variety of grape, the soil, the weather all influence the outcome. If you want to truly know whether your hood is safe, don't settle for one lens. Look through both. And remember that in Boston, Berlin, or Beijing, chemistry doesn't care which standard you follow. Fumes will escape if the hood doesn't contain them.

Our job, no matter which flag flies above the building, is to make sure they don't.

Chapter 19:
Compliance
Checking the Box or Ensuring Safety?

When the word "compliance" comes up in a lab setting, most people breathe a sigh of relief. We're compliant, they say, as if that alone settles the question of safety. But does it?

For many organizations, fume hood testing is treated as a compliance exercise - a box to be checked, a report to file away in case OSHA, the EPA, or an accrediting body ever asks for it. The assumption is simple: if the paperwork says the hoods were tested, then the hoods must be safe. But reality, as we have learned again and again, is not so tidy.

Imagine I hand you five ASHRAE 110 test reports, each from a different manufacturer's hood. Could you, by looking at the numbers alone, tell me which hood is the best and why?

Chances are, you could not. The reports might all show passing results, but those results are snapshots, not a full picture of performance. The hood that "complies" today may fail tomorrow, and the hood that looks good on paper may leak vapor into the breathing zone when someone walks by or when the building's air supply shifts.

That is the gap between compliance and safety. Compliance is about meeting the minimum requirements of a regulation. Safety is about ensuring that people aren't being exposed to hazardous chemicals. Those two goals overlap, but they are not the same thing.

The Compliance Trap
Standards like **ASHRAE 110** and **ANSI/ASSP Z9.5** were never meant to be the end of the conversation. ASHRAE 110 is a method of testing, not a performance guarantee. Z9.5 calls its provisions minimum criteria and encourages innovation so long as exposure is controlled. OSHA, too, requires that hoods function properly, but it does not define "safe" solely by a face velocity number.
Yet in practice, compliance often gets reduced to one or two easy-to-measure items:

- Did we measure face velocity?
- Did the report get filed?

The problem is that face velocity alone tells us little about containment. A hood can show "100 feet per minute" and still spill contaminants into the lab due to turbulence, cross drafts, or poor work practices. Studies confirm that 15–30% of hoods in service do not provide adequate containment, even though most of them are technically "compliant".

When Compliance Fails
Let's take a real scenario. A university lab proudly shows that every hood passed annual certification. The reports all look neat, with averages in the "safe" range. Six months later, a graduate student smells solvent vapors while working. A Tri-Color laser visualization test is brought in, and suddenly the picture changes: plumes of haze leak out every time someone walks past the hood.

So, what went wrong? Nothing - if you measure success only by compliance. Everything - if your true goal is safety.

This is why relying on compliance alone can create a dangerous illusion. It can make managers, EHS staff, and users believe they are safe when in fact exposures are still happening.

The Shift in Thinking

Compliance should not be the goal. It should be the byproduct of a hood that is functioning safely. A well-designed, properly installed, and regularly challenged hood will naturally comply with the regulations. But a hood that merely complies does not always protect its user.

The better question is not "Is my hood compliant?" but rather "Is my hood protecting me from exposure?" That is why performance-based testing - tracer gas, dynamic visualization, containment challenges - are so important. They move us past the checkbox and closer to real assurance.

Compliance as the Floor, Not the Ceiling

We must remember what these standards represent. Z9.5 calls its requirements "minimum criteria". ASHRAE 110 provides a reproducible method of testing but explicitly states it does not define safe procedures. OSHA's General Duty Clause requires employers to protect workers from recognized hazards, whether or not a specific regulation exists.

The message is clear: compliance is the floor, not the ceiling. True safety requires going beyond the bare minimum - challenging assumptions, testing dynamically, and focusing on containment.

The Story We Should Tell

The story of compliance in the lab world is not one of villains, but of habits. We check boxes because it is easy, because the form looks official, and because regulators ask for it. But safety is not a form. Safety is what happens - or doesn't happen - when chemicals are released in front of a graduate student, a technician, or a chemist with decades of experience.

The day we stop asking "Is it compliant?" and start asking "Is it safe?" is the day compliance finally finds its rightful place: not as the goal, but as the side effect of doing things right.

Chapter 20:
Legal Liability
The Hidden Risk No One Wants to Talk About

When we talk about fume hoods, the conversation usually centers on airflow, face velocity, or the latest testing protocols. But there's a quieter, heavier subject that hangs over every EHS officer, lab manager, and tester: **legal liability**.

Let's imagine for a moment that there is an accident. A chemical release. Someone is injured-or worse, killed. Investigators will show up. Lawyers will get involved. And very quickly, everyone connected to that fume hood-the facility, the safety officer, the tester, the manufacturer, even outside consultants-could find themselves drawn into the legal spotlight.

It doesn't matter if you were directly responsible or just part of the system. In the United States, litigation is broad and inclusive. If your name is on the paperwork, if you signed the test report, if you were the designated "responsible person," chances are good that you'll be named in the lawsuit. Even if your role was minor, you may be dragged into depositions, discovery, and potentially years of legal proceedings.

I've often said: safety is invisible until it fails. Unfortunately, liability works the same way. Everything looks fine-until it isn't.

The Reality of Shared Responsibility

The laboratory environment is a web of overlapping duties. OSHA makes it clear that employers have a responsibility to provide a workplace "free from recognized hazards". Z9.5 and other standards lay out expectations for laboratory ventilation, recordkeeping, and responsible persons. SEFA and ASHRAE provide guidance on exposure control devices.

But in a courtroom, all those standards and guidelines can become weapons. A plaintiff's attorney may hold up a test report and ask:

- Why didn't you retest after that failure?
- Why wasn't there training when you knew room conditions mattered?
- Why didn't you follow the hierarchy of controls outlined in the standards?

Even if the accident was primarily caused by poor work practices, the paper trail will be used to show negligence or inadequate oversight.

Professional Protection

That's why I tell every EHS manager, every industrial hygienist, and every tester: don't just think about airflow-think about insurance. Professional liability insurance is not optional; it's essential. It may feel unnecessary when things are running smoothly, but when the worst happens, it may be the only thing standing between you and financial ruin.

And remember, insurance isn't just about you personally. Institutions should ensure that liability coverage extends to the teams responsible for laboratory ventilation and exposure control devices. Because in the legal world, the net is always cast wide.

Real-World Examples of Liability in Action

The UCLA Sheri Sangji Case (2008)
When UCLA research assistant Sheri Sangji died from severe burns after a pyrophoric chemical fire, the tragedy made national headlines. But what shocked the academic world even more was the legal aftermath: UCLA and the supervising professor were both charged with felony violations of California labor laws. This was one of the first times a university and a PI faced criminal charges for a lab accident. The case highlighted that liability doesn't stop at the institutional level-it can reach individuals in supervisory and safety roles.

Texas Tech University Explosion (2010)
In another case, a graduate student at Texas Tech was severely injured in a laboratory explosion. The Chemical Safety Board (CSB) investigation revealed lapses in hazard recognition, training, and oversight. While no one was killed, the legal exposure was massive. The university faced OSHA scrutiny, potential civil suits, and reputational damage that lasted for years.

Dow Chemical (Midland, MI)
In industry, too, liability is real. Dow and other chemical companies have faced lawsuits over worker exposures tied to containment failures and unsafe laboratory practices. Even when multiple safeguards existed, the question in court was always: was everything reasonably possible done to prevent harm?
Each of these cases serves as a warning. The responsibility chain is long, and when something goes wrong, it gets pulled tight.

Lessons From the Field

I've seen it firsthand. After a hood incident, blame doesn't stop at the lab bench. It travels upstream-to the lab manager, to the EHS officer, to the facilities team, to the external certifier who signed the last test report. Even when those people did nothing wrong, they were forced to defend themselves, often at great cost in time, money, and stress.

Here's the hard truth: if you are connected to fume hood performance, you are connected to liability.

A Final Word

Testing hoods is about safety-but it's also about **legal defense**. Your reports may one day be Exhibit A in a courtroom. The procedures you followed may be scrutinized by people who know nothing about fume hoods but everything about building a case.

That's why it's not enough to meet the minimums. It's not enough to say, "we complied." Compliance is the floor, not the ceiling. True safety-and real liability protection-comes from demonstrating that you did everything reasonable to ensure containment, protect users, and mitigate risks.

Test thoroughly. Document carefully. Train continuously. And protect yourself with professional liability coverage. Because when the courtroom lights go on, the question won't be just "was the hood safe?" but also "what did you do to make sure?"

If you are interested in the costs of accidents liability, we have a book dedicated to the subject.

Chapter 21: The Hidden Danger of LEL When Slow Airflow Turns Deadly

We spend much of our time in laboratories thinking about exposure limits-how to keep concentrations of chemicals below OSHA's Permissible Exposure Limits (PELs) or the ACGIH TLVs. But there's another set of limits that doesn't get nearly enough attention: the **explosive limits**.

Flammable gases, vapors, and some dusts that can form explosive mixtures There is a lower explosive limit (LEL) and an upper explosive limit (UEL). If the vapor concentration falls between those two boundaries and an ignition source is present, you no longer have a safe workspace-you have the conditions for an explosion. It doesn't matter if the source is a hot plate, a spark from static electricity, or even the quiet clicking of a stir plate motor. Once that concentration window is reached, the lab is a tinderbox.

🔥 Substances That Have an LEL

LEL applies to **combustible materials in the gas or vapor phase**, including:

- **Hydrocarbons** - methane, propane, butane, hexane, benzene, toluene, etc.
- **Alcohols** - methanol, ethanol, isopropanol, etc.
- **Solvent vapors** - acetone, ether, xylene, etc.
- **Industrial gases** - hydrogen, ammonia, acetylene, carbon monoxide, etc.
- **Vapors from fuels** - gasoline, diesel, jet fuel, etc.

- **Certain dusts** (under special testing) - grain, sugar, coal dust, metal powders (though these are usually characterized by minimum explosible concentration rather than LEL).

Now, how does this tie back to fume hoods?

The Vortex No One Talks About

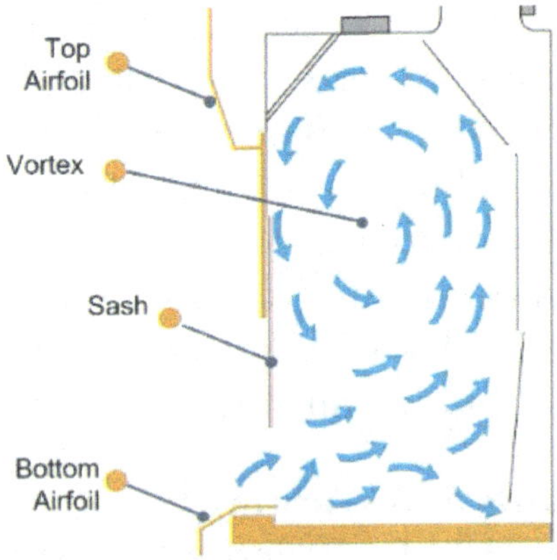

Traditional fume hood designs often create a **vortex in the upper portion of the chamber**. This rotating eddy of air sits just behind the sash and moves up and down as the sash is raised or lowered. In theory, exhaust air should sweep it clean, but in practice many hoods are poor at flushing out this pocket. What does that mean? Any flammable vapors released inside the hood can migrate into this stagnant swirl and linger there. Over time, instead of being safely diluted and exhausted, they can become concentrated.

This is especially true when we've slowed the face velocity. In recent decades, labs have pushed for lower velocities (60–80 fpm instead of 100–120 fpm) to save energy and reduce turbulence but it comes with a dark side. **Slower airflow means slower dilution.** And in the case of flammable vapors, that's a dangerous trade-off.

From Invisible Accumulation to Flash Fire

Imagine a researcher transferring diethyl ether inside a hood. The vapor is heavier than air and highly flammable. As the transfer continues, vapors rise into the vortex. With reduced face velocity and poor design, there isn't enough flushing of the vortex to dilute it. Slowly, silently, the concentration in that pocket edges upward.

Below the LEL, there's no problem. But once the vapor concentration crosses that line, say 1.9% for ether, the conditions are set. One spark, one static discharge, and the vortex becomes a flame ball. That flame front doesn't politely stay put-it bursts outward, blowing through the sash opening and into the room.
We've documented lab fires where this very scenario is believed to have happened. Most were called "mystery flashes," but the airflow dynamics tell the real story.

Standards and Warnings
Both **ASHRAE 110** and **ANSI/ASSP Z9.5** emphasize that fume hoods are exposure control devices-not explosion-proof chambers. The OSHA Laboratory Standard (29 CFR 1910.1450) further reminds employers that hoods provide containment, not guaranteed fire protection. In fact, most manufacturers explicitly state in their manuals that hoods are not designed for handling explosives or uncontrolled quantities of flammable vapors.

Most hood designers pay little attention to vortex. Especially how well it flushes. A hood design that allows buildups in the vortex is an accident waiting to happen.

Over the past 20 years we have seen many more fume hood fires and explosions. In many cases we can only speculate the root cause, but since the function of the hood is to capture, contain, dilute and exhaust when there is less air there is less dilution. This is where safety and sustainability collide. Poor hood design wasn't as much of an issue when there was more air.

Word of caution

As we consider testing hoods with IPA or Propylene both having a LEL we could be creating a hazard in a poorly designed hood where the vortex isn't flushing effectively.

Seeing the Invisible

Modern visualization tools like **Tri-Color laser testing** have shown how vortices trap and circulate vapors. When you watch haze climb into the upper chamber and stay there, it's not hard to imagine flammable vapors doing the same thing. And unlike haze, vapors don't just disappear-they accumulate until chemistry takes over.

The Balancing Act

So how do we reconcile these competing forces-energy savings versus explosion safety?

- **Risk Assessments First**: Any procedure involving significant flammable vapor release should trigger a review. Is the hood appropriate? Is additional dilution air or a specialty hood needed?

- **Velocity Matters**: Research shows extremely low face velocities (<60 fpm) or excessively high ones (>150 fpm) both compromise performance. Somewhere in the middle lies the sweet spot where containment is strong enough to flush vapors but not so fast as to create turbulence.

- **Don't Trust the Average**: A hood may show 80 fpm average, but if that vortex isn't being cleared, it's irrelevant. Containment tests and visualization are the only way to know.
- **Training the User**: Lab staff must understand that hoods are not explosion-proof boxes. Working with flammables requires smaller quantities, minimized open transfers, and careful housekeeping to avoid dangerous buildup.

A Lesson in Limits

The story of LEL inside a fume hood is one of unintended consequences. In our quest for quieter, more efficient, greener labs, we risk overlooking the explosive potential of what we can't see.

Face velocity isn't just about exposure-it's also about **dilution safety**. And if we ignore the invisible vortex, we may learn the hard way that when air slows down, chemistry speeds up

This flash fire/explosion was caused by igniting 3 drops of Isopropyl Alcohol (IPA) in this container.

Chapter 22:
Myths and Misconceptions
What Everyone Gets Wrong About Fume Hoods

Walk into almost any laboratory and you'll hear the same refrain: "If the face velocity is at 100 feet per minute, the hood is safe." It's repeated so often it has become a kind of mantra, passed from professor to student, technician to manager, as though carved into stone tablets. But like many myths, it's built on half-truths and oversimplifications. And those myths are dangerous, because they give a false sense of security.

Let's start with that sacred cow of fume hood performance: face velocity. For decades, 100 feet per minute (fpm) was treated as the golden number. Too low, and fumes would escape. Too high, and turbulence would blow vapors back into the room. The truth, however, is more complicated. Research has shown that face velocity alone has little correlation with containment. A hood running at 100 fpm might still leak vapors because of turbulence, cross-drafts, or poor design. Conversely, another hood might protect users perfectly well at 60 fpm if the airflow patterns are stable and smooth. In other words: airflow quality matters more than airflow quantity.

Another common misconception is that compliance equals safety. Many institutions test their hoods once a year, get a passing mark and check the box. The hood must be safe, right? Not necessarily.

Field experience shows that about half of all hoods fail containment testing even when their face velocities appear acceptable. Failures often trace back to three culprits: poor hood or room design, ventilation system imbalances, or unsafe user practices. A report may satisfy compliance, but it doesn't guarantee protection. Compliance should be the byproduct of safety-not the substitute for it.

Then there's the dangerous belief that "more is better." If 100 fpm is good, then 150 fpm must be great, right? Wrong. Too much airflow can cause turbulence, whipping vapors out of the hood and into the user's breathing zone. Think of air as a river. A steady, laminar current carries everything downstream. But hit the rapids, and eddies form, splashing water back onto the rocks. The same principle applies in a hood: excessive velocity creates chaos instead of containment.

Another myth that persists is that the fume hood is a standalone device. Many users think, "As long as I'm working behind the sash, I'm protected." The reality is that a hood is just the front end of a much larger system-the laboratory ventilation system. If the supply air registers blow too forcefully across the hood face, or the exhaust system can't maintain negative pressure, the best-designed hood will still fail. Hoods don't work in isolation; they are part of a delicate dance between room pressurization, supply air, and exhaust air.

And perhaps the most dangerous misconception of all: "If I can't see it, it's not there." Vapors are invisible, odorless, and silent. For decades, users assumed that if no smells leaked out, the hood was doing its job. But containment failures can occur long before you notice any odor. That's why tools like tracer gas testing and Tri-Color laser visualization are so important-they make the invisible visible. With lasers slicing across the hood opening, you can see exactly when and where vapors escape. It's often a sobering sight for researchers who assumed their hood was flawless.

So why do these myths persist? Partly because they're convenient. A single number-like face velocity-offers a simple answer in a world where people crave certainty. Compliance reports provide a piece of paper to wave at regulators. And because most exposures don't result in immediate injury, the risks remain invisible until a catastrophic accident or long-term health issue forces people to pay attention.
But the lesson is clear: don't be lulled into a false sense of security by myths and shortcuts. A fume hood is not safe just because a sticker says so, or because the velocity meter reads 100 fpm. True safety comes from a holistic approach: regular and meaningful testing, good room design, proper training, and an understanding of how people actually work at the hood.

The fume hood should be your laboratory's most trusted ally against chemical exposure. But only if you learn to separate myth from reality. As the old saying goes, "What you don't know can hurt you." And in the lab, what you can't see can kill you.

Conclusion
If We're Not Willing to Do It Right

So far, we've explored nearly every aspect of fume hood performance testing—from face velocity and tracer gas to visualization, standards, and user behavior. My hope is that you now see fume hood testing not as a routine compliance task, but as a critical safeguard for human life. Maybe you've started to think about things you hadn't before—the invisible forces, the myths we've long accepted, and the simple truth that numbers alone can't prove containment.

Every fume hood tells a story, and every story has a backstory. The history of how we test, what we measure, and why we measure it has shaped the way laboratories operate today. But somewhere along the way, context was lost. Urban myth replaced original intent, and many of us began chasing face velocity numbers instead of focusing on what truly matters—containment.

If we want to keep labs and the people who work in them safe, our mission must be simple and unwavering: **focus on containment.** The only way to know whether a hood contains safely is to test it—properly, thoughtfully, and under real conditions.

We need to eliminate the "lick-'em-stick-'ems," those quick sticker tests that give the illusion of safety. Fume hood testing should be easy and economical, yes—but it must also be safe, meaningful, and environmentally responsible.

We must remember that the user is not a bystander but an active part of the ventilation system. Testing is not just verification; it's education. Each test, each visualization, each discussion is a training opportunity—a chance to help someone understand how their behavior impacts containment and, ultimately, their own safety.

If we are not willing to do it right—if we're content with shortcuts, guesswork, or false assurances—then we might as well turn the hoods off and save the energy. But if we are willing to do it right, to test with purpose, to train with passion, and to lead with integrity—then together, we can make laboratories safer, one fume hood at a time.

And If We're Not Willing to Do It Right, maybe we should just shut off the hoods and save the energy. That way the user will know the hood isn't working.

Free Fume Hood Stickers for Safety Training

If you would like some of these labels reach out to
experts@fumehoodcertified.com

Get Hoodie Gear

https://fumehoodtesterscatalog.com/collections/hoodie-gear

Take advantage of Fume Hood Certifier's extensive resources!

fumehood.academy
Information regarding our in-person workshops in Phoenix. A great event for any experience level.

fumehoodcertified.com
Learn about all that FHC has to offer and how we may able to help you

tricolorasertesting.com
Home of the Tri-Color Visualizer system. Learn how we utilize a modified ASHRAE110 large volume test to challenge fume hood's dynamic containment abilities.

fumehoodonlinecourse.com
Chip's last book, "Laboratory Fume Hoods Explained" converted into a free online course

fumehoodperformance.com
The home for all your fume hood testing data for easy sharing, review and analysis. Works seamlessly with our Q200 and Face Velocity Profiler data capture software

fumehoodsafety.com
We host several introductory level videos here for students and employees that are new to fume hoods to get a understanding of safe user practices

fumehoodsoftware.com
Learn all about our Fume Hood Performance Tracker here, includes a walk-through video and a live demo.

fumehoodtesterscatalog.com
A one stop shop for everything you need to perform ASHRAE110 compliant fume hood testing, from smoke testing to Q200 factory calibrations, we have it all!

fumehoodtestingnetwork.com
Directory of industry professionals that utilize our Tri-Color testing system.

Safety is No Accident (Blog)
https://fumehoodtesterscatalog.com/blogs/lab-safety-digest

YouTube
www.youtube.com/@FumeHoodCertified

LinkedIn
www.linkedin.com/company/fume-hood-certified-llc

Facebook
www.facebook.com/fumehoodcertified

Laboratory Fume Hoods Explained (Free Online Course)
www.fumehoodonlinecourse.com

Bulk Sales

Help Spread the word.

We are offering large discounts for multiple copies. Reach out to us at experts@fumehoodcertified.com

Do you have a Medical Portal?

Just like people use medical portals to monitor their health, laboratory fume hoods now have their own health tracker- **Fume Hood Performance Tracker**.

It's your digital portal to ensure every fume hood is operating safely and efficiently.

Fume Hood Certified is offering a **free trial** to help you get started. Discover how easy it is to track performance, flag issues before they become problems, and document safety like never before.

Don't wait for a safety incident to start monitoring your hoods-be proactive.

Email us today at **experts@fumehoodcertified.com** and start your free trial of the Fume Hood Performance Tracker.

HOOD IDENTIFICATION NUMBER

23-R24 Y44

SCAN ME

FHC FUME HOOD CERTIFIED

Your lab's safety deserves the same attention as your own health. Let us help you make that possible.

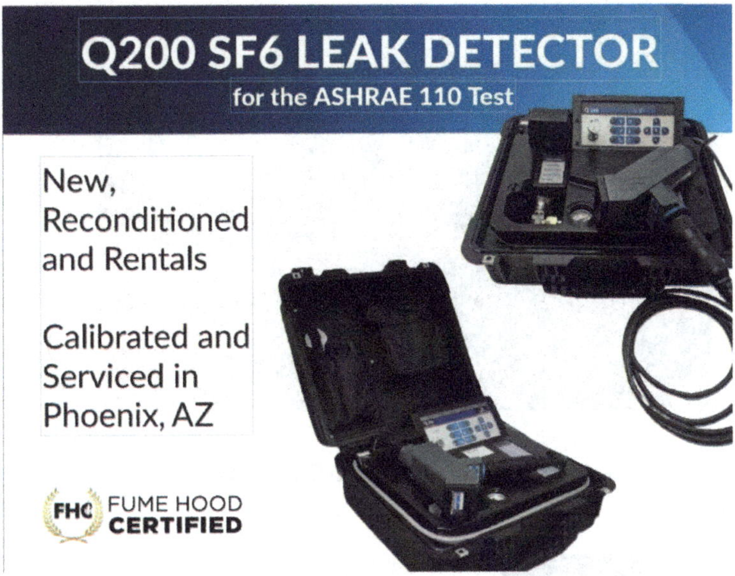
Q200 SF6 LEAK DETECTOR
for the ASHRAE 110 Test
New, Reconditioned and Rentals
Calibrated and Serviced in Phoenix, AZ
FHC FUME HOOD CERTIFIED

DO YOU WANT TO BECOME A CERTIFIED FUME HOOD TESTER??
CERTIFIED FUME HOOD TESTER
FHC FUME HOOD CERTIFIED
CAN HELP!
FULLY ONLINE!
ASHRAE 110 - 2016 Tester Training & Certification

www.fumehood.academy

THE CONTAINMENT CHRONICLES

www.fumehoodcertified.com/fhc-newsletter

www.fumehoodtesterscatalog.com

Made in United States
Orlando, FL
04 November 2026

90202248R00066